IEG WORLD BANK
INDEPENDENT EVALUATION GROUP

I0126236

Climate Change and the World Bank Group

Phase I: An Evaluation of World Bank Win-Win Energy Policy Reforms

2009
The World Bank
Washington, D.C.

http://www.worldbank.org/ieg

World Bank InfoShop
E-mail: pic@worldbank.org
Telephone: 202-458-5454
Facsimile: 202-522-1500

Independent Evaluation Group
Knowledge Programs and Evaluation Capacity
Development (IEGKE)
E-mail: ieg@worldbank.org
Telephone: 202-458-4497
Facsimile: 202-522-3125

Printed on Recycled Paper

Contents

Tables

A visible spectral band image of Tropical Storm Hudah, April 2000. Photo courtesy of the Visible Earth Team, NASA.

Abbreviations and Terminology

bcm	Billion cubic meters
CAS	Country Assistance Strategy
CDM	Clean Development Mechanism
CEA	Country Environmental Analysis
CFL	Compact fluorescent light bulb
CO_2	Carbon dioxide
CO_2e	Carbon dioxide equivalent
DPL	Development Policy Loan
DSM	Demand-side management
EER	Energy-Environment Review
ERR	Economic rate of return
ESCO	Energy service company
ESMAP	Energy Sector Management Assistance Program
gas	Natural gas
gCO_2	Grams of carbon dioxide
GDP	Gross domestic product
GEF	Global Environment Facility
GHG	Greenhouse gas
GGFR	Global Gas Flaring Reduction Partnership
Gt	Billion tons
GTZ	German Technical Cooperation
GW	Gigawatt
IBRD	International Bank for Reconstruction and Development (World Bank)
IDA	International Development Association
IEA	International Energy Agency
IEG	Independent Evaluation Group
IFC	International Finance Corporation
IMF	International Monetary Fund
kg	Kilogram
kW	Kilowatt
kWh	Kilowatt-hour
LNG	Liquefied natural gas
mmbtu	Millions of British thermal units
mscf	Thousand standard cubic feet
MW	Megawatt
NO_x	Nitrogen oxides
OECD	Organisation for Economic Co-operation and Development
OPEC	Organization of Petroleum Exporting Countries
PCF	Prototype Carbon Fund
PER	Public Expenditure Review
ppm	Parts per million

PSIA	Poverty and Social Impact Analysis
REDD	Reduced Emissions from Deforestation and Degradation
SEA	Strategic Environmental Analysis
SO_2	Sulfur dioxide
SO_x	Sulfur oxides
tCO_2e	Tons CO_2 equivalent
ton	Metric ton (=tonne; 1,000 kg)
TW	Terawatt
UNFCCC	United Nations Framework Convention on Climate Change

Glossary

Adaptation	Measures taken by societies and individuals to adapt to actual or expected adverse impacts on the environment, especially as the result of climate change.
Biodiversity	Short for biological diversity. Refers to the wealth of ecosystems in the biosphere, of species within ecosystems, and of genetic information within populations.
Carbon capture and storage	A technology for preventing the release of carbon dioxide to the atmosphere from thermal power plants by capturing the gas and storing it underground.
Carbon dioxide equivalent (CO₂e)	A standard unit for measuring the impact of a greenhouse gas on global warming. For instance, one ton of methane is considered equivalent in warming to 25 tons of carbon dioxide.
Carbon accounting (and/or carbon footprint)	Measurement of the gross or net impact on greenhouse gas emissions of an organization, project, or program.
Carbon fund	A fund set up for the purchase of carbon credits.
Carbon offset (or credit)	A financial instrument representing a reduction in greenhouse gas emissions (including gases other than carbon dioxide), used by purchasers to meet regulatory or voluntary limits on emissions.
Carbon shadow pricing	The practice of incorporating into the economic analysis of projects or programs an economic value associated with the external costs of greenhouse gas emissions or external benefits of emissions reduction.
Certified emission reduction	A carbon credit (measured in tons CO₂e) for an emissions reduction associated with a Clean Development Mechanism project.
Clean Development Mechanism	"A mechanism under the Kyoto Protocol through which developed countries may finance greenhouse-gas emission reduction or removal projects in developing countries, and receive credits for doing so which they may apply towards meeting mandatory limits on their own emissions" (UNFCCC).
Climate change	Changes in climatic conditions and processes (including but not limited to warming) that go beyond natural climatic variability. When used in connection with mitigation, refers to human-induced changes.

Combined-cycle turbine	A relatively efficient technology for power generation from combustion, usually of natural gas.
Demand-side management	Actions or incentives, often directed by energy utilities to their customers, to reduce the level of energy demands (typically through efficiency measures) or change the timing of those demands.
District heating	Centralized system for the provision of steam heat to an urban neighborhood or district.
Ecosystem	The interacting system of a biological community and its nonliving environmental surroundings.
Emission	In this volume, emission primarily refers to the anthropogenic release of greenhouse gases, as from fossil fuel combustion or deforestation. Used also to refer to other kinds of air pollution from combustion, such as particulates and sulfur oxides.
Energy services company	A company that provides clients with some combination of assessment, financing, and implementation of options for increased efficiency of use and reduced expenditure on energy.
Environment	The sum of all external conditions affecting the life, development, and survival of an organism.
Environmental assessment	A process whose breadth, depth, and type of analysis depend on the proposed project. It evaluates a project's potential environmental risks and impacts in its area of influence and identifies ways of improving project design and implementation by preventing, minimizing, mitigating, or compensating for adverse environmental impacts and by enhancing positive impacts.
Environmental impact	Any change to the environment, whether adverse or beneficial, wholly or partially resulting from an organization's activities, products, or services (as defined in ISO 14001).
Environmental mainstreaming	The integration of environmental concerns into macroeconomic and sectoral interventions.
Environmental sustainability	Ensuring that the overall productivity of accumulated human and physical capital resulting from development actions more than compensates for the direct or indirect loss or degradation of the environment. Goal 7 of the UN Millennium Development Goals specifically refers to this, in part, as integrating the principles of sustainable development into country policies and programs and reversing loss of environmental resources.
Gas flaring	Burning of natural gas, usually when released as an unintended by-product of oil production.

Greenhouse gas	Gases whose atmospheric buildup contributes to global warming and climate change. Greenhouse gases regulated under the Kyoto Protocol are carbon dioxide, methane, nitrous oxide, hydrofluorocarbons, perfluorocarbons, and sulphur hexafluoride.
Mitigation	Measures taken to reduce adverse impacts on the environment.
Netback price	Wellhead value of natural gas computed by netting transport costs from final market price.
Ozone-depleting substances	Manufactured chemical compounds that reduce the protective layer of ozone in the Earth's atmosphere. The Montreal Protocol, administered by the UN, maintains the list of ozone-depleting substances that are targeted for control, reduction, or phase-out.
Performance Standards	The eight Performance Standards establish requirements that the client is to meet in IFC-financed projects.
Safeguard policies	Policies designed specifically to ensure that the environmental and social impacts of projects supported by the Bank Group are considered during appraisal and preparation. The Bank's safeguard policies cover environmental assessment, natural habitats, pest management, indigenous peoples, cultural resources, involuntary resettlement, forests, dam safety, international waterways, and disputed areas.
Sustainable development	Development that meets the needs of the present without compromising the ability of future generations to meet their own needs.
Win-win policy	Here, a policy that provides net benefits both to the nation that adopts it and to the world at large. Individuals or groups may suffer losses under win-win policies, though in principle they could be compensated from the benefits. Also called no-regrets policy.

The cooling towers of an old power plant in Soweto are no longer in use. One now depicts local art and the other advertises a local power company, Photo by Christian Schlaeger, reproduced with his permission.

Acknowledgments

Kenneth Chomitz was the evaluation manager and main author for this study. Major contributions to chapter 5 on efficiency policies were made by Meredydd Evans and Bin Shui.

The evaluation also drew on background studies and evaluative work by Charles Ebinger (power policies), Donald Hertzmark (natural gas), and Craig Meisner (cross-national analyses of energy consumption and power sector fuel mix). Principal research assistants Dinara Akhmetova, Ashwin Bhouraskar, and Kunal Khatri undertook diligent portfolio analysis. Victoria Gunnarson, Stephen Hutton, Romain Lacombe, Urvashi Narain, and Yadviga Semikolenova also provided valuable assistance.

Peer reviewers Fernando Manibog, Siv Tokle, and David Wheeler and external panel reviewers Geoffrey Heal and Thomas Heller provided useful feedback on the evaluation draft. The panel reviewers, including Rajendra K. Pachauri, also provided comments on the final draft that will also guide the next phase of the evaluation.

Initial drafts of the report benefited from editing by William Hurlbut; the report was edited for publication by Caroline McEuen with assistance from Heather Dittbrenner. Nik Harvey assisted in publication and managed Web site production. Gloria Mestre-Soria and Nischint Bhatnagar provided administrative support. Vivian Jackson, Alex McKenzie, and Melanie Zipperer assisted in dissemination. Thanks go to Ismail Arslan, Arup Banerji, Sharokh Fardoust, Ali Khadr, and many others at IEG for advice and help.

The evaluation team is grateful to David Victor and colleagues at Stanford for discussions and notes on the political economy of power reform. The team is also grateful for the cooperation of World Bank staff members and others who were interviewed.

IEG gratefully acknowledges InWEnt's cosponsorship of a workshop related to Phase II of the evaluation series.

Director-General, Evaluation: *Vinod Thomas*
Director, IEG-World Bank: *Cheryl Gray*
(director at inception: *Ajay Chhibber*)
Task Manager: *Kenneth M. Chomitz*

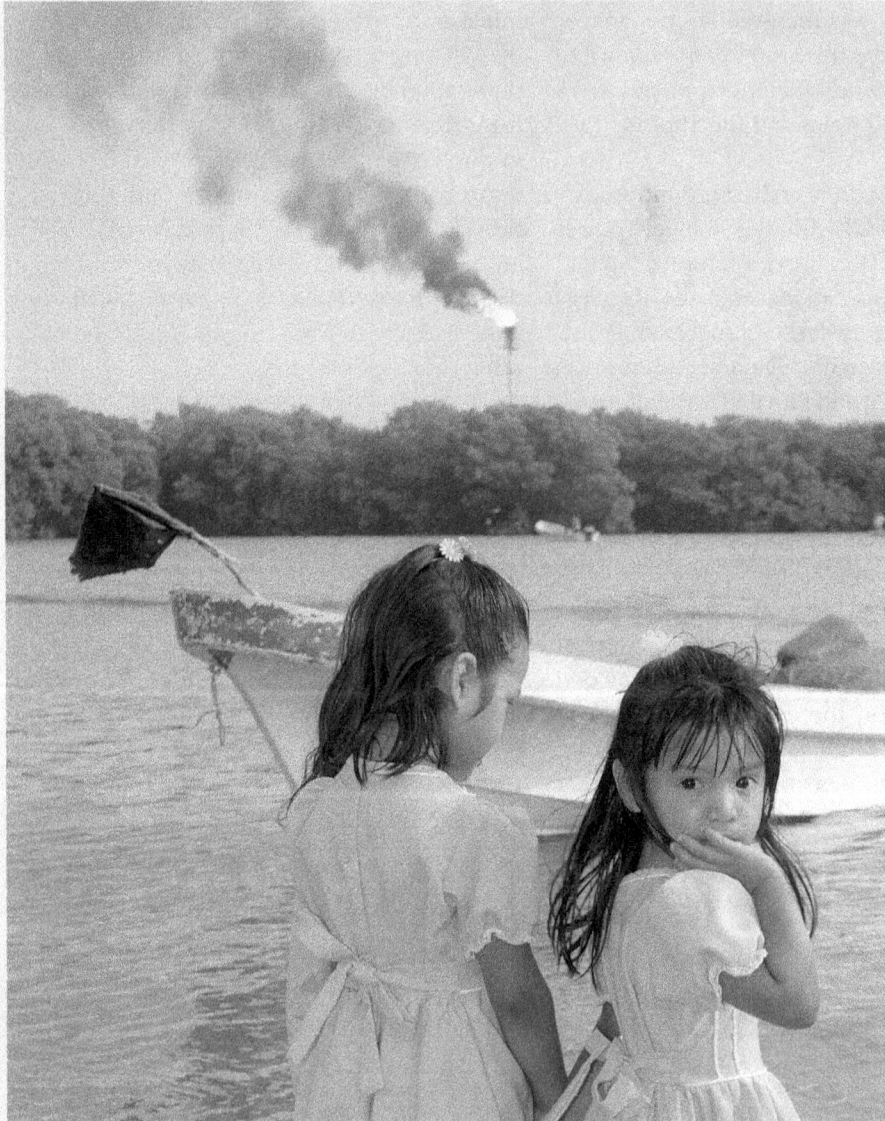

A natural gas flaring tower at Pemex's Dos Bocas petroleum-exporting complex, Mexico. Reproduced by permission of Corbis; photo © Keith Dannemiller/Corbis.

Foreword

Scientific consensus warns that climate change threatens to derail development, while business-as-usual development threatens to destabilize the climate. The World Bank Group has awakened to the challenge of disarming these interlocking risks. But in doing so, it has to confront areas of possible tension:

- Between a country-focused operational model and support for global public goods
- Between a global role encompassing developed countries and its focus on developing nations
- Among greenhouse gas mitigation, climate adaptation, and near-term growth.

Win-win policies in energy pricing and in non-price energy efficiency have the potential to reconcile national and global goals. They can help countries meet a good part of their incremental energy needs at low cost, while freeing up funds for social protection and increasing resilience to international energy price shocks. About a fifth of the baseline global increase in energy-related CO_2 emissions could be reduced by 2030 through efficiency measures that pay for themselves, in the developing world alone.

Policy reforms are needed to unlock these benefits. Energy price reform is seldom easy, but 2008 market conditions showed the unsustainability of energy subsidies, and the Bank is well placed to help. Analytic and financial support can promote socially beneficial and politically feasible options—for instance, redirection of poorly targeted energy subsidies to social protec-

tion. The Bank's investments in energy efficiency have often been effective, but they have been modest, with little emphasis on policies. There is change, however, including a recent ramp-up in International Finance Corporation investments. Countries are receptive, and Bank Group leadership could make a difference to this up-to-now under-prioritized area.

Win-win policies will not be enough to meet clients' energy needs or to decouple development from emissions. The UN Framework Convention on Climate Change stresses developed countries' responsibilities to reduce their own emissions and to provide financial and technological support to developing countries. Relevant to this is Bank Group experience in using concessional and carbon finance to support clean energy technologies—the subject of the second phase of the climate evaluation. IEG is also assessing forest sector experience that bears on reduced emissions from deforestation.

The Bank has had limited direct experience in adaptation, although efforts in disaster prevention and weather index insurance are cases that suggest consonance with near-term development goals. Adaptation is the subject of the climate evaluation's third phase.

The World Bank Group has a vital role in addressing the interlinked problems of development and climate change. IEG's three-year program of evaluation is designed to assist the Bank Group as it formulates and implements an operational strategy in this critical area.

Vinod Thomas
Director-General, Evaluation

Foreword

A coal-run power plant in Tangshan, China, in China's Hebei Province. Reproduced by permission from Corbis; photo © Jason Lee/Reuters/Corbis.

Executive Summary

Climate change threatens to derail development, even as development pumps ever-greater quantities of carbon dioxide into an atmosphere already polluted with two centuries of Western emissions. The World Bank, with a newly-articulated Strategic Framework on Development and Climate Change, must confront these entangled threats in helping its clients to carve out a sustainable growth path.

But this is known territory—many of the climate change policies under discussion have close analogues in the past. This phase of the evaluation, focused on the World Bank (and not the International Finance Corporation or the Multilateral Investment Guarantee Agency), assesses the World Bank's experience with key win-win *policies* in the energy sector—policies that combine gains at the country level with globally beneficial greenhouse gas (GHG) reductions. The next phase will look across the entire World Bank Group at *project-level* experience in promoting technologies for renewable energy and energy efficiency and at some issues related to climate change in the Bank's transport and forestry portfolios.

Within the range of win-win policies, this report examines two that have long been discussed but are more relevant than ever in light of record energy prices: removal of energy subsidies and promotion of end-user energy efficiency. Energy subsidies are expensive, damage the climate, and disproportionately benefit the well-off. Their reduction can encourage energy efficiency, increase the attractiveness of renewable energy, and allow more resources to flow to poor people and to investments in cleaner power. Though subsidy reduction is never easy, the Bank has a record of accomplishment in this area, especially in the transition countries. About a quarter of Bank energy projects included attention to price reform.

Improvements in the design and implementation of social safety nets can help to rationalize energy prices while protecting the poor.

End-user energy efficiency has long been viewed as a win-win approach with great potential for reducing emissions. It becomes increasingly attractive as the costs of constructing and fueling power plants rise. About 5 percent of the Bank's energy commitments *by value* (about 10 percent *by number*) have gone to specific efficiency efforts, including end-user efficiency and district heating. Including a broader range of projects identified by management as supporting supply-side energy efficiency would boost the proportion above 20 percent by number. Few projects tackled regulatory issues related to end-user efficiency, though the Bank has invested in some technical assistance and analytical work. This historical lack of emphasis on energy efficiency is not unique to the Bank and reflects the complexity of pursuing end-user efficiency, a pervasive set of biases that favor electricity supply over efficiency, inadequate investments in learning, and inattention to energy systems in the wake of power sector reform.

The record levels of energy prices in 2008, although they have been relaxed, provide an impetus for the Bank and its clients to choose more sustainable long-term trajectories of growth. The mid-2008 oil price was equivalent to the 2006 price, plus a $135 per ton tax on carbon dioxide—the kind of level that energy modelers say is necessary for long-term climate stabilization. To help clients cope with the burden of these prices, and take advantage of the signals they send for sustainability, the Bank can do four things:

1. It can make promotion of energy efficiency a priority, using efficiency investments and policies to adjust to higher prices and constructing economies that are more resilient.
2. It can assist countries in removing subsidies by helping to design and finance programs that protect the poor and help others adjust to higher prices.
3. It can promote a systems approach to energy.
4. And it can motivate and inform these actions, internally and externally, by supporting better measurement of energy use, expenditures, and impacts.

Goals and Scope

This evaluation is the first of a series that seeks lessons from the World Bank Group's experience on how to carve out a sustainable growth path. The World Bank Group has never had an explicit corporate strategy on climate change against which evaluative assessments could be made. However, a premise of this evaluation series is that many of the climate-oriented policies and investments under discussion have close analogues in the past, and thus can be assessed, whether or not they were explicitly oriented to climate change mitigation.

This report, which introduces the series, focuses on the World Bank (International Bank for Reconstruction and Development and International Development Association), and not on the International Finance Corporation (IFC) or the Multilateral Investment Guarantee Agency (MIGA). It assesses its experience with key win-win *policies* in the energy sector: removal of energy subsidies and promotion of end-user energy efficiency. The next phase looks at the expanding project-level experience of the Bank and the IFC in promoting technologies for renewable energy and energy efficiency; it also addresses the role of carbon finance. A parallel study examines the role of forests in climate mitigation. The climate evaluation's final phase will look at adaptation to climate change.

Motivation

Operationally, the World Bank has pursued three broad lines of action in promoting the mitigation of GHG emissions, the main contributor to climate change. First, it has mobilized concessional finance from the Global Environment Facility (GEF) and carbon finance from the Clean Development Mechanism (CDM) to promote renewable energy and other GHG-reducing activities. Second, and to a much more limited extent, it has used GEF funds to stimulate the development of noncommercial technologies. Third, and the subject of this evaluation, it has supported win-win policies and projects—sometimes with an explicit climate motivation, often without. These actions not only provide global benefits in reducing GHGs, but also pay for themselves in purely domestic side benefits such as reduced fuel expenditure or improved air quality. The win-win designation obscures the costs that these policies may impose on particular groups, even while benefiting a nation as a whole. This presents challenges for design and implementation.

Two sets of win-win policies are perennial topics of discussion in the energy sector: reduction in subsidies and energy-efficiency policies, particularly those relating to end-user efficiency. This report looks at these, and at another apparently win-win topic: gas flaring. Flaring is interesting because of its magnitude, the links to pricing policy and to carbon finance, and the existence of a World Bank–led initiative to reduce flaring.

Findings

Development spurs emissions.

A 1 percent increase in per capita income induces—on average and with exceptions—a 1 percent increase in GHG emissions. Hence, to the extent that the World Bank is successful in supporting broad-based growth, it will aggravate climate change.

But there is no significant trade-off between climate change mitigation and energy access for the poorest.

Basic electricity services for the world's unconnected households, under the most unfavorable assumptions, would add only a third of a percent to global GHG emissions, and much less if renewable energy and efficient light bulbs could be deployed. The welfare benefits of electricity access are on the order of $0.50 to $1 per kilowatt-hour, while a stringent valuation of the corresponding carbon damages, in a worst-case scenario, is a few cents per kilowatt-hour.

Country policies can shape a low-carbon growth path.

Although there is a strong link between per capita income and energy-related GHG emissions, there is a sevenfold variation between the most and least emissions-intensive countries at a given income level. Reliance on hydropower is part of the story behind these differences, but fuel pricing is another. High subsidizers—those whose diesel prices are less than half the world market rate—emit about twice as much per capita as other countries with similar income levels. And countries with long-standing fuel taxes, such as the United Kingdom, have evolved more energy-efficient transport and land use.

Energy subsidies are large, burdensome, regressive, and damage the climate.

The International Energy Agency's 2005 estimate of a quarter-trillion dollars in subsidies each year outside the Organisation for Economic Co-operation and Development (OECD) may understate the current situation. While poor people receive some of these benefits, overall the benefits are skewed to wealthier groups and often dwarf more progressive public expenditure. Fuel subsidies alone are 2 to 7.5 times as large as public spending on health in Bangladesh, Ecuador, the Arab Republic of Egypt, India, Indonesia, Morocco, Pakistan, Turkmenistan, República Bolivariana de Venezuela, and the Republic of Yemen. At the same time, subsidies encourage inefficient, carbon-intensive use of energy and build constituencies for this inefficiency.

The Bank has supported more than 250 operations for energy pricing reform.

Success has been achieved in the transition countries—in Romania and Ukraine, for example, where energy prices were adjusted toward market levels, and the intensity of carbon dioxide emissions dropped substantially. Subsidy removal can threaten the poor, however. Recent efforts to assess poverty and welfare impacts systematically appear to have informed the design and implementation of price reform efforts, though not necessarily with direct Bank involvement. Examples include Ghana and Indonesia, where compensatory measures were deployed in connection with fuel price rises.

The Bank has rarely coordinated efficiency improvements with subsidy reductions to lighten the immediate adjustment burden on energy users.

An exception is the China Heat Reform and Building Efficiency Project, which links improved insulation with heat pricing. A growing number of projects sponsor nationwide distribution of compact fluorescent light bulbs, but this has been done in response to power shortages (Rwanda, Uganda) or to stanch utility losses (Argentina, Vietnam), rather than to facilitate subsidy reduction.

Despite emphasis on energy efficiency in Bank statements and in Country Assistance Strategies (CASs), the volume and policy orientation of IBRD/IDA efficiency lending has been modest.

Although the IFC has recently increased its investments in energy-efficiency projects, World Bank commitments for efficiency were about 5 percent by value of energy finance over 1991–2007. This includes investments in demand-side efficiency and district heating, and may also include some supply-side efficiency investments. By this definition, about 1 in 10 projects by number involve energy efficiency. Including a broader range of projects identified by management as supporting supply-side energy efficiency would boost the proportion above 20 percent by number over the period 1998–2007. Globally only about 34 projects undertaken over the 1996–2007 period had components oriented to demand-side energy-efficiency policy. Among these, many attempts to promote efficiency have had limited success because the Bank has engaged with utilities, which have limited incentives to restrict electricity sales.

There are several reasons why end-user energy-efficiency projects, and especially policy-oriented projects, appear to be under-emphasized in the Bank's portfolio.

The Bank has carried out some successful and innovative efficiency projects. But internal Bank incentives work against these projects because they are often small in scale, demanding of staff time and preparation funds, and may require persistent client engagement over a period of years. There is a general tendency to prefer investments in power generation, which are visible and easily understood, over investments in efficiency, which are less visible, involve human behavior rather than electrical engineering, and whose efficacy is harder to measure. A general neglect of rigorous monitoring and evaluation reinforces the negative view of efficiency.

The Bank-hosted Global Gas Flaring Reduction Partnership (GGFR) has fostered dialogue on gas flaring, but it is difficult to assess its impact on flaring activity to date.

Associated gas (a by-product of oil production) is often wastefully vented or flared, adding more than 400 million tons of carbon dioxide equivalent to the atmosphere annually, or about 1 percent of global emissions. A modestly funded public-private partnership, the GGFR has succeeded in highlighting the issue, promoting dialogue, securing agreement on a voluntary standard for flaring reduction, and sponsoring useful diagnostic studies. But only four member countries have adopted the standard. The GGFR has emphasized carbon finance as a remedy for flaring, but the use of project-level carbon finance is a mere bandage for policy ailments that require a more fundamental cure.

Recommendations

In mid-2008, real energy prices were at a record high. While this is burdensome for energy users, it opens an opportunity for the Bank to support clients in making a transition to a long-term sustainable growth path that is resilient to energy price volatility, entails less local environmental damage, and is a nationally appropriate contribution to global mitigation efforts.

Clearly the World Bank needs to focus its efforts strategically on areas of its comparative advantage. This would include supporting the provision of public goods and promoting policy and institutional reform at the country level. Furthermore, the Bank can achieve the greatest leverage by promoting policies that catalyze private sector investments in renewable energy and energy efficiency, including those supported by IFC and MIGA.

The analysis in this report supports the following recommendations:

Systematically promote the removal of energy subsidies, easing social and political economy concerns by providing technical assistance and policy advice to help reforming client countries find effective solutions, and analytical work demonstrating the cost and distributional impact of removal of such subsidies and of building effective, broad-based safety nets.

Energy price reform can endanger poor people and arouse the opposition of groups used to low prices, thereby posing political risks. But failure to reform can be worse, diverting public funds from investments that fight poverty and fostering an inefficient economy increasingly exposed to energy shocks. And reform need not be undertaken overnight. The Bank can provide assistance in charting and financing adjustment paths that are politically, socially, and environmentally sustainable. Factoring political economy into the design of reforms and supporting better-targeted, more effective social protection systems will be elements of this approach.

Emphasize policies that induce improvement in energy efficiency as a way of reducing the burden of the transition to market-based energy prices.

Historically, energy efficiency has received rhetorical support but garnered only a small share of financial support or policy attention. This is beginning to change with such moves as China's commitment to drastically reduce its energy intensity and India's Energy Conservation Act. But the Bank can do much more to help clients pursue this agenda. If a real reorientation to energy efficiency and renewable energy is to occur, the Bank's internal incentive system needs to be reshaped. Instead of targeting dollar growth in lending for energy efficiency (which may skew effort away from the high-leverage, low-cost interventions), it needs to find indicators that more directly reflect energy savings and harness them to country strategies and project decisions. It needs also to patiently support longer, more staff-intensive analysis and technical assistance activities. Increased funding for preparation, policy dialogue, analysis, and technical assistance is required.

Promote a systems approach by providing incentives to address climate change issues through cross-sectoral approaches and teams at the country level, and structured interaction between the Energy and Environment Sector Boards.

To tackle problems of climate change mitigation and adaptation, the Bank and its clients need to think, organize, and act beyond the facility level, and outside subsectoral and sectoral confines. One avenue for this is through greater attention to systemwide energy planning. Integrated resource planning, once in vogue, has been largely abandoned in the wake of power sector privatization and unbundling. Yet current planning methods are inadequate in integrating considerations of end-use efficiency and in balancing the risks of volatile fuel prices and weather-sensitive electricity output from wind and hydropower

plants. Water management, urban management, and social safety nets are other areas where cross-sectoral collaboration is essential to promoting win-win policies and programs.

Invest more in improving metrics and monitoring for motivation and learning—at the global, country, and project levels.

Good information can motivate and guide action.

First, building on the Bank's current collaboration with the International Energy Agency on energy efficiency indicators, the Bank could set up an Energy Scoreboard that will regularly compile up-to-date standardized information on energy prices, collection rates, subsidies, policies, and performance data at the national, subnational, and project levels. Borrowers could use indicators for benchmarking; in the design and implementation of country strategies, including sectoral and cross-sectoral policies; and in assessing Bank performance.

Second, more rigorous economic and environmental assessment is needed for energy investments and those that release or prevent carbon emissions. These assessments should draw on energy prices collected for the Scoreboard; account for externalities, including the net impact on GHG emissions; and account for price volatility. Investment projects should also be assessed, qualitatively, on a diffusion index, which would indicate the expected catalytic effect of the investment in subsequent similar projects. It is desirable to complement project-based analysis with assessment of indirect and policy-related impacts, which could be much larger.

Third, monitoring and evaluation of energy interventions continue to need more attention. Large-scale distribution of compact fluorescent light bulbs is one example of an intervention that is well suited to impact analysis and where a timely analysis could be important in informing massive scale-up activities.

Rising waters threaten a cement plant in Bangladesh. Photo by Jouni Martti Eerikainen, reproduced with his permission.

Management Response

M anagement welcomes the evaluation by the Independent Evaluation Group (IEG) of some of the World Bank's experience with "win-win" energy policy reforms, which constitute an important but not exhaustive set of activities within the wider suite of World Bank Group efforts on the energy front.

It is useful to take stock of progress on the win-win reforms as defined by IEG, as they are an important element of the World Bank Group's vision to contribute to inclusive and sustainable globalization—to help reduce poverty, enhance growth with care for the environment, and expand individual opportunity. In this context, management particularly would welcome the promised second phase of IEG's evaluation, covering the expanding project-level experience of the Bank and International Finance Corporation (IFC) in promoting renewable energy, energy efficiency, and carbon finance, the absence of which precludes a comprehensive assessment of the focus and success of World Bank Group efforts on the energy front.

Overview of Response

Management concurs with several aspects of IEG's main findings, many of which reinforce important messages already captured in the Bank's energy sector practices or in the findings from Bank economic and sector work, internal reviews and self-evaluation, and emerging lessons from operational experience across the World Bank Group. At the same time, management takes issue with the evaluation scope of IEG's report; its definition of win-win energy opportunities; the gaps in evaluated areas; and the use, in certain cases, of findings to draw overly broad conclusions or recommendations, such as promoting the use of integrated resource planning by regulators of supply-side energy entities. Therefore, in several respects, management differs with IEG's findings and recommendations.

Key Issues of Agreement and Divergence

This management response first outlines the areas in which management broadly agrees with the analysis in the review, noting, however, areas where IEG could have given a fuller account of efforts the World Bank has made or is making. It then discusses areas in which management believes that IEG has drawn conclusions from an analysis based on limited coverage or that do not fully take into account the underlying context.

Areas of Agreement

Management agrees with the importance of energy efficiency and energy pricing in the Bank's work and the need for strong collaboration across sectors on energy policy issues. However, management believes that the report does not adequately reflect the considerable work the Bank has undertaken to address energy efficiency. The Bank's strong involvement in energy efficiency began in the late 1970s/early 1980s in response to oil price shocks. Although interest in energy efficiency languished after the subsequent fall in oil prices, it was rekindled in the early 1990s when Eastern European and former Soviet Union countries became active borrowers. During the 1990s, the Bank supported energy efficiency reforms in Europe and Central Asia Region countries through a combination of technical assistance, policy loans, and investment projects.[1] The role of energy efficiency was further reinforced by the Bank's *Fuel for Thought* (World Bank 2000), which pushed for market-based approaches to energy efficiency.

Post-Bonn Efforts. The World Bank Group has followed up on its commitment made at the 2004 Bonn International Conference on Renewable Energy to increase annual energy efficiency and new renewable energy lending by 20 percent, starting in fiscal year 2005. Indeed, average fiscal 2005–07 energy-efficiency commitments have more than doubled compared with the previous three-year period. The World Bank continues to scale up energy efficiency work in the energy sector. Staffing up to increase the skills base is well under way in both the anchor and operational units. Energy efficiency specialists have been/are being hired by Regional units, Carbon Finance, and the Energy Sector Management Assistance Program (ESMAP).

Areas of Divergence

Management believes that IEG has drawn conclusions from an analysis based on limited coverage or that do not fully take into account the underlying context. Management is concerned that limitations on both definitions and the scope of IEG's report open the way to mischaracterization of the extent and impact of World Bank Group effort on energy efficiency.

Circumscribed Scope. The evaluation scope of IEG's report is circumscribed, incorporating only International Bank for Reconstruction and Development (IBRD) and International Development Association (IDA) energy-efficiency policy, energy pricing, and gas flaring initiatives, while excluding IFC's substantive role (except, very occasionally, at the margins). Management observes that excluding IFC programs and activities that target the key private sector role in promoting energy efficiency is a major shortcoming. IFC activities encompass a range of initiatives (such as the Efficient Lighting Initiative) and sustainability advisory services. By focusing piecemeal on Bank policy experience and deferring project-level experience to a second phase of review, IEG has not taken into account that the efforts of each of the World Bank Group's components are intended to complement one another and build on respective comparative advantages and synergies, and it has precluded a comprehensive evaluation of the energy efficiency experience in the World Bank Group. As a result, management observes that some of the report's Phase 1 findings paint an incomplete picture of World Bank and World Bank Group efforts on the energy front.

Definition of Win-Win. IEG's report uses a narrow definition of win-win energy opportunities. Management is concerned that the report focuses on, and draws conclusions from, one dimension of energy efficiency (end-user energy efficiency), while not adequately incorporating other important win-win energy opportunities, in particular, supply-side energy efficiency (which covers power plant rehabilitation to improve efficiency and also electricity transmission and distribution system loss reduction), renewable energy, and fuel switching.

Indicator. The IEG report uses an indicator that is limited to "specific efficiency efforts, including end-user efficiency and district heating." This opens the way to conclusions and perceptions that may be misleading, including that only about 1 in 10 World Bank energy projects involves energy efficiency. However, as noted in the IEG report, "including a broader range of projects identified by management as supporting supply-side energy efficiency would boost the proportion above 20 percent by number."[2]

Management, and certainly the clients of the World Bank Group, would have benefited from a more comprehensive analysis and an indicator that included all energy supply-side efficiency, technical assistance, and development policy lending, as well as IFC investments in energy efficiency.

Management Action Record. Management's specific responses to IEG recommendations are outlined in the attached draft Management Action Record.

Management Action Record

Recommendation	Management Response
Systematically promote the removal of energy subsidies, easing social and political economy concerns by providing technical assistance and policy advice to help reforming client countries find effective solutions, and analytical work demonstrating the cost and distributional impact of removal of such subsidies and of building effective, broad-based safety nets.	***Agreed; work is already ongoing.***
Energy price reform, never easy or painless, can pose social and political economy risks in client countries. But the Bank can help provoke and promote reforms by providing clients with assistance in charting and financing adjustment paths that are politically, socially, and environmentally sustainable.	The Bank continues to work with client countries to address the issue of energy subsidies. Technical assistance and policy advice are provided, as requested by our client countries. The Bank focuses on the legal and regulatory mechanisms needed to support sustainable energy pricing reforms.
One way to do this is for the Bank to continue to develop and share knowledge on the use of cash transfer systems or other social protection programs as potentially superior alternatives to fuel subsidies in assisting the poor. This would include systematic analyses of the distributional impact of energy subsidies. Timely monitoring and analysis of energy use and expenditure, at the household and firm levels, will also be important in policy design, in securing public support, and in detecting and repairing holes in the safety net.	Energy staff will continue to work with Poverty Reduction and Economic Management Network and Human Development Network staff (for example, *Guidance for Responses from the Human Development Sectors to Rising Food and Fuel Prices,* World Bank HDN 2008) to develop and apply social safety nets, including cash transfers, designed to protect the poor from the impact of energy price adjustments. A regulatory thematic group has been established in the Bank to foster dissemination of lessons learned. These lessons will be applied, taking into account the unique circumstances in client countries. When requested, the Bank provides support to enable countries to monitor and analyze energy use so that findings can be applied to their energy policies.
Emphasize policies that induce improvement in energy efficiency as a way of reducing the burden of transition to market-based energy prices.	***Partially agreed; work is already ongoing.***
Cost-reflective prices for energy boost the returns to efficiency, but the Bank should support country policies that allow households and firms to exploit efficiency opportunities. Conversely, the deployment of energy-efficient equipment such as compact fluorescent lights can be used as a device for cushioning the impact of price increases. The Bank should explore innovative ways to finance efficiency (and renewable energy) investments in the face of fuel price volatility.	The Bank has established an Energy Efficiency for Sustainable Development program to help guide and scale up energy efficiency activities. It is implementing the first step of this program, to increase the staffing with energy-efficiency experience, in ESMAP, the Energy Anchor Unit, and the Regions. This effort is complemented by a learning program developed by the Bank's energy-efficiency thematic group, under the oversight of the Energy and Mining Sector Board. Another step is the development of programs and projects at the country/policy level, the industry level, and the equipment level to ensure that a broad-based implementation program evolves.

Management Action Record

Recommendation	Management Response
	To foster World Bank Group support for energy efficiency, the draft "Development and Climate Change: A Strategic Framework for the World Bank" (World Bank 2008) has proposed an initiative to screen the project pipeline for energy-efficiency potential early in the project design phase.
	The Bank is working with the donor community to: (i) increase the financial support needed to intensify energy-efficiency efforts; (ii) increase low-cost funding to support energy-efficiency and renewable energy programs; and (iii) broaden the support from partners in implementing a renewable energy and energy-efficiency program.
In order to strengthen internal incentives toward promotion of energy efficiency, the Bank should develop appropriate metrics, such as indicators that more directly reflect energy savings, instead of dollar growth targets in lending for energy efficiency (which may distort effort away from the high-leverage, low-cost interventions). These indicators, in turn, need to be harnessed to country strategies and project decisions. All of these efforts are likely to call for increased funding for preparation, policy dialogue, analysis, and technical assistance rather than lending.	In terms of internal incentives, the discussion on developing appropriate metrics has been ongoing with the International Energy Agency and with UN Energy, but to date it has been inconclusive. Given the inconclusive nature of the discussion to date, management is not prepared to agree with establishing new metrics that focus solely on energy efficiency. The World Bank Group has committed to accelerate lending for new renewable energy and energy efficiency to 30 percent per annum over the next three years, a 50 percent increase over the 2004 Bonn commitment (which it has consistently met since that time).
Promote a systems approach by providing incentives to address climate change issues through cross-sectoral approaches, teams at the country level, and structured interaction between the Energy and Environment Sector Boards.	*Partially agreed; work is already ongoing.*
Helping clients reform will require a systems view, such as looking at the power system as a whole; looking at energy subsidies as just one, undesirable, part of a social protection system; and looking at the connections between water and power management.	The Bank will continue to use a system-wide approach in reviewing projects and programs.
To be effective the Bank needs to break down sectoral silos and encourage cross-sector approaches and teams. This will require championship by country directors and vice presidents, to promote incentives such as supporting capacity building for power system regulators in integrated resource planning, and using the Clean Technology Fund to support public systems that will catalyze widespread investments.	Most Regions and many country teams have already created climate change teams of staff from several sectors to promote synergies, and are developing cross-sectoral business strategies to integrate climate change considerations. The World Bank Group established a Climate Change Management Group as a focal point to discuss cross-sectoral issues and promote synergies. The Bank supports regulatory capacity building, drawing on les-

Management Action Record

Recommendation	Management Response
	sons learned from successful cases accomplished to date. On the basis of previous experience, management disagrees with the proposed use of integrated resource planning, as it is unconvinced of the effectiveness of the use of integrated resource planning by either supply-side entities or their regulators.
	However, the Bank supports the use of broad-based planning tools by policy makers to support the implementation of policies in the legal and regulatory framework.
	The Bank is currently considering large-scale responses to demand-side issues using new funding for low-carbon technologies when the funds become available.
Structured interaction of the Energy and Environment Sector Boards, initiated with ad hoc groups to address specific cross-sectoral challenges, could move the Bank closer toward mainstreaming sustainable development.	The merging of infrastructure and environment into a common vice presidency has facilitated interaction at the sector boards and thematic working groups.
Invest more in improving metrics and monitoring for motivation and learning at the global, country, and project levels.	*Partially agreed; work is already ongoing.*
Good information can motivate and guide action. One particularly useful global initiative for the World Bank would be to collaborate with the International Energy Agency or other partners to set up an Energy Scorecard that would compile up-to-date and regular standardized information on efficiency indicators, energy prices, policies, and subsidies at the national and sectoral levels. Indicators could be used by borrowers for benchmarking; in the design and implementation of country strategies, including sectoral and cross-sectoral policies; and in assessing Bank performance in assisting countries.	The Bank has been working with the International Energy Agency on collecting energy-efficiency–related information in pilot countries for two years, with limited success. Management does not commit to the idea of establishing a centrally maintained Energy Scorecard. Rather, the focus of our efforts is now on helping client countries establish their capacity to undertake the data collection exercise in a manner that targets both effective implementation and related policy-making guidance. Without this capacity and country willingness to participate in and lead this initiative, it will not be sustained. The Bank is also looking into possible new, innovative knowledge-sharing mechanisms to facilitate sharing lessons learned.
At the national level, the Bank should support integration of household and firm surveys with energy consumption and access information to lay the foundation for assessing impacts of price rises and mitigatory measures, as well as planning for improved access.	The Bank lacks the resources to maintain a comprehensive and reliable database on energy policies, prices, subsidies, and energy efficiency at the national level. Regional organizations provide part of this information, which the Bank selectively draws upon, depending on the information's reliability.

Management Action Record	
Recommendation	**Management Response**
	The Bank, with ESMAP support, has led in improving Living Standards Measurement Survey (LSMS) instruments for increased collection of energy data as part of LSMS surveys.
At the project level, the Bank should invest in rapid-feedback monitoring and impact evaluation of efficiency projects and policies.	The Bank will include rapid-feedback and monitoring and impact evaluation of efficiency projects when requested by our borrowers.

Chairperson's Summary: Committee on Development Effectiveness (CODE)

On August 27, 2008, the Committee on Development Effectiveness (CODE) met to consider the report entitled *Climate Change and the World Bank Group—Phase I: An Evaluation of World Bank Win-Win Energy Policy Reform* prepared by the Independent Evaluation Group (IEG), together with the draft Management Response.

Background

On December 17, 2007, the Committee considered a study entitled *The Welfare Impact of Rural Electrification: A Reassessment of the Costs and Benefits*, prepared by IEG. The Committee considered the IEG report *Supporting Environmental Sustainability—An Evaluation of World Bank Group Experience, 1990–2007*, and draft Management Response on June 18, 2008. Recently, the Committee discussed the draft *Strategic Framework on Climate Change for the World Bank Group* at its meeting of August 6, 2008.

IEG Evaluation

IEG introduced the current evaluation report as part of a phased series on climate change. Subsequent phases will address issues of clean technology investments, carbon finance, and adaptation, and will look across the World Bank Group. This Phase I evaluation assessed the World Bank's experience with key win-win policies in the energy sector—those that combine gains at the country level with globally beneficial greenhouse gas (GHG) reductions. The analysis of this report supported the following recommendations:

- Systematically promote the removal of energy subsidies, easing social and political economy concerns by providing technical assistance and policy advice to help reforming client countries find effective, broad-based safety nets.

- Emphasize policies that induce improvements in energy efficiency as a way of reducing the burden of transition to market-based energy prices.

- Promote a systems approach by providing incentives to address climate change issues through cross-sectoral approaches and teams at the country level and structured interaction between the energy and environment sector boards.

- Invest more in improving metrics and monitoring for motivation and learning at the global, country, and project levels.

Draft Management Response

Management agreed with the importance of energy efficiency and energy pricing in the Bank's work and the need for collaboration across sectors on energy policy issues. At the same time, management believes that IEG has drawn conclusions from an incomplete analysis based on limited coverage and that do not fully take into account the underlying context. Management expressed concerns that the IEG report does not cover the full range of the World Bank Group's programs and activities (for example, assisting the private sector in promoting energy efficiency) and that it focuses on one subset of win-win energy opportunities and excludes others, such as energy conservation, load management, and supply-side efficiency investments, as well as renewable energies and fuel switching.

Overall Conclusions

The Committee commended IEG for an excellent report, which members found very informative, and acknowledged the trade-offs of undertaking the evaluation in appropriate, sequenced parts as had been outlined and agreed in the Approach Paper. Nevertheless, it was essential that strategic communication be carefully designed to avoid misleading or unfair interpretations of the findings. The plan for a capstone paper covering all three phases was endorsed. There was strong support for deepening the Bank's engagement with clients on energy pricing policies, though there was recognition that it is a complex issue encompassing economic, environmental, social, and political aspects that were likely to vary country by country and over time. The Bank could play a useful role in sharing best practices and distilling lessons of experience, particularly on energy taxes and subsidies and on pricing policies for renewable energy to help countries institute socially and environmentally sustainable pricing.

The general sentiment was for greater emphasis than hitherto on energy pricing policy, and energy efficiency in a broad sense. In this regard, the issues of external institutional incentives and internal incentives resonated with several attendees who recommended that management pay greater attention to this matter, including one suggestion to consider organizational changes (noting parenthetically that this issue's relevance goes well beyond the energy sector). While noting management's point about dividing labor appropriately with other agencies such as the International Energy Agency (IEA), the broad sentiment at the meeting was supportive of IEG's recommendations that the Bank be more involved in developing metrics and performance indicators. Indeed, several speakers added that

analytical and design work in this regard should be at a global level, encompassing developed countries as well. Thus, the World Bank Group could play a very useful role in making high-quality information and a balanced monitoring framework for a global public good.

Next Steps

The report is the first of a three-part IEG evaluation on Climate Change and the World Bank Group, and focuses on IBRD-IDA experience. In response to the Committee's request, IEG committed to clarify the scope, content, and context of the Phase I report as part of its preparation for publication. This includes clarifying how it fits in the three-phase evaluation by IEG (where the second phase will look at the World Bank Group's project-level experience in promoting technologies for renewable energy, energy efficiency, and transport; and the third phase will look at adaptation issues). IEG also committed to prepare a capstone paper summarizing the three phases at the conclusion of the series; the Committee will consider whether or not to recommend this paper for a full Board discussion.

Main Issues Raised at the Meeting

The principal issues discussed were the following:

Scope of IEG Report

Some speakers would have liked to have seen immediate treatment (in the current phase) of a broader range of topics, including energy conservation and energy access; supply-side in addition to demand-side efficiency; discussion of new and additional financing, particularly for technology and equipment; discussion of additional energy sources, including biofuel or nuclear; coverage and targeted analysis of Bank support for adaptation; and extension of the evaluation beyond energy to forestry, transport, and agriculture issues. One member agreed with IEG's recommendations but felt that further thought should be given on how to implement them.

IEG's definition of win-win (or no-regret) policies and projects offering potential gains at the country level aligned to global interest (for example, reduction in GHG) drew some comments. One member felt the report could have expanded this concept to consider environmental taxation and subsidies for renewable energy. Some others underscored that the paper should have given more emphasis to the principle of "common but differentiated responsibilities and respective capacities" in emissions and in additional financing, rather than focusing on savings from removal of subsidies. In this regard, a member noted that the poorest countries, which emit only a tiny fraction of the per capita emissions of developed countries, will be disproportionately affected by climate change. At the same time, the need to address subsidy reductions and energy efficiency in developed countries was raised by another speaker.

Some members stressed the importance of broadening the evaluation to World Bank Group activities, including synergies between institutions. One speaker considered that the structure of IEG's proposed suite of climate-related analyses would be incomplete without explicitly addressing the GHG implications of the Bank Group's engagements to help developing countries reform their power sectors. This speaker suggested that IEG should evaluate the positive and negative links between different power sector reforms and low-carbon electricity services as part of the second phase of its climate evaluation. IEG said that Phase I focused mainly on the World Bank, but the next phase will certainly include the International Finance Corporation and the Multilateral Investment Guarantee Agency. A few members suggested an appropriate communication strategy for disseminating the IEG three-phased review in a comprehensive manner to avoid misunderstandings. As suggested by some speakers, IEG agreed to highlight, during the dissemination of each phase of the report, that it is part of a broader review.

Bank's Assistance

The Bank was encouraged to deepen its engagement with countries through policy dialogue and to support them to pursue appropriate regulatory and institutional settings. Some speakers stressed the importance of adjusting the internal (for staff and management) and external (countries, Bank, and development partners) institutional incentive system. However, they also cautioned about the need to consider political economy considerations, as well as market failure and institutional constraints in client countries. A question was raised about the adequacy of the Bank's resources as well as organizational and operational capabilities to address the challenges of policy dialogue and reforms. In addition, one member stressed the need to balance the emphasis between software (price reform and regulatory framework) and hardware (energy-efficiency equipment). Management affirmed the Bank's internal capacity to provide a full package: 200 experts in thematic teams and cross-sectoral teams in the Regions, offering not only lending but also technical assistance, as well as social safety nets and policy advice.

Subsidies and Energy Pricing

There was general consensus on the need to be mindful of the political challenges of subsidies and pricing reforms, as well as economic and social dimensions at the national and regional levels. Speakers agreed that more emphasis should be given to removal of energy subsidies and were not surprised by IEG findings that subsidies were a poorly monitored drag on the economies of developing countries. They also stressed the importance of supporting energy pricing reform, an area recommended by IEG for greater emphasis. On price reform, the importance of diversity of reform packages to address country-specific circumstances; of a gradual approach to complement progress in institutional development; of finding windows of opportunity for analytical work and policy dialogue to motivate reform; and of client ownership were noted. It was also added that the adjustment of prices to market level should take into account vulnerable groups in relation to the other interests vested in the society, and the need for appropriate compensation systems.

Speakers encouraged the Bank to disseminate lessons learned, good practices, and guidelines, as well as more analytic work on implementing various reforms including fiscal sustainability, cross-subsidization, distributional impact, and cap-and-trade schemes. Management indicated that the Bank uses a number of instruments to appreciate the political economy, such as Poverty and Social Impact Analyses. Management also noted that the Organisation for Economic Co-operation and Development (OECD) has done work on best practices in environmental taxation and cap-and-trade that the Bank is using in its analysis. Some speakers stressed the importance of addressing energy subsidies analysis and energy pricing reform in the new Strategic Framework on Climate Change and Development (SFCCD), which management indicated would be addressed in the full SFCCD paper.

Efficiency Policies

Some speakers agreed with IEG on the need for the Bank to systematically encourage more energy-efficiency activities in client countries. Management agreed, and stated that the full range of interventions, including the supply side of energy efficiency (loss reduction in distribution, transmission, and generation), and alternatives such as buses and public transportation systems need to be taken into account, depending on the country-specific circumstances. While acknowledging the importance of supply-side efficiency, IEG stressed that demand-side efficiency measures have been viewed by recent studies as offering the largest opportunities for energy savings and emissions reductions—larger than those offered by supply-side measures. Demand-side and end-use efficiency require policy attention because of underlying market failures

and have been repeatedly stressed in Bank policy documents.

Metrics and Monitoring

Several speakers concurred with IEG's recommendation that the Bank should work toward developing appropriate metrics, while recognizing management's point that data collection would be costly. A few speakers pointed to a 1999 ESMAP "scorecard" publication as precedent. Additionally, some speakers stressed the need for the Bank to play an advocacy role in promoting a more balanced global monitoring mechanism by including indicators such as mobilizing financial and technological support to developing countries, while the political sensitivities and technical complexities of carbon accounting were acknowledged. Management indicated that it does not commit to developing and maintaining a database of this type, but it will work to develop indicators and help countries to establish capacity. Management noted that the Bank works together with the OECD, EUROSTAT, and multilateral development banks, and supports specialized agencies such as the IEA and UN, trying to help them formulate better indicators.

Global Gas Flaring Reduction Partnership (GGFR)

A few speakers noted that the Bank has played an advocacy role in promoting reduction of gas flaring, but that adherence to the initiative has been below expectations. Questions were raised on whether there was a lack of interaction between the GGFR and Bank's business or lack of competitiveness of the Bank's financial instruments.

Jiayi Zou, Chairperson

Statements by the External Review Panel: Climate Evaluation, Phase I

Geoffrey M. Heal
Paul Garrett Professor of Public Policy and Business Responsibility, Columbia University

Overall I think this is a very good report. It focuses on important issues that are ones where the Bank can make some difference. My comments are minor.

I think that the two main themes, removal of energy subsidies and improvement of energy efficiency, are critical issues in the context of developing countries (and rich countries too!) facing rising energy prices and threatened by climate change. We know from experience that neither is easy to achieve, but for both I feel sure that the benefits outweigh the costs and fully justify the efforts. I do think it is particularly important to stress, as the report does, that removing energy subsidies need not compromise the ability to get energy to the poorest in society more efficiently, and that the main beneficiaries of subsidies are often the middle and upper classes. I was struck by the numbers indicating that high subsidizers have much higher emissions per capita than others: not surprising, but the numbers are impressive.

The report refers several times in the early sections to a systems approach to energy. I am still not completely sure what is meant by this. I take it to mean looking simultaneously at all aspects of energy production and consumption and thinking through interactions and possible duplication and overlap, worrying more about joint heat and power schemes, and so on. It is likely that there are real gains in this area but I feel that this is something that should be spelled out more clearly.

I was impressed by the comment that the social benefits of providing power to the poorest greatly outweigh the social costs, even if power is provided in a way that generates greenhouse gases. These numbers should be more widely known. They are important in the global discussions on climate change and the role of the poor countries in mitigating this.

I like the suggestion of Energy Scorecards. These can provide a basis for benchmarking, often important in the policy-making context, and could also be useful in climate negotiations. Connected to this is the idea of carbon pricing of projects that emit CO_2, even when there is no legal requirement to purchase permits. Most major banks in the West now require this of their clients: U.S. banks, for example, require their clients to charge for carbon emissions in project evaluations even though there is no need to buy carbon permits. It would be natural for the Bank to do this too.

As the report mentions, emissions from deforestation are large and generated by developing countries: Brazil, Indonesia, and China are in the top four emitters, and for Brazil and Indonesia it is the case that most emissions come from deforestation. There is scope for a global win-win move if we implement one of the Reduced Emissions from Deforestation and Degradation (REDD) ideas now under discussion, as this will not only reduce emissions but also lead to new development finance. The Bank's Prototype Carbon Fund is important in this context.

Again, in summary, I was impressed by the review: it seems to address very important issues, and does so clearly.

Thomas C. Heller
Lewis Talbot and Nadine Hearn Shelton Professor of International Legal Studies, Stanford University

My comments are intended to be useful and provocative, even though I understand that, as detailed in chapter 1, the segment of the overall projected IEG evaluation we have before us is very restricted. It deals with win-win opportunities and defers systematic consideration of major issues (like carbon markets) that are only alluded to in this initial treatment. Any criticism of findings or recommendations in these areas of work key to rating and reforming Bank Group performance is evidently unfair as premature. Still, I hope that these remarks on the incomplete work may contribute to shaping the entire final product.

I want to state immediately that I like the report and find its organization, analyses, and recommendations generally clear, well founded, and pertinent. I will describe below the main points that exemplify these contributions. After stressing my strong appreciation for the tenor and content the report already makes (part A), I would like to discuss an implicit issue that runs throughout that is troubling (part B). The issue is that even a cursory history of the Bank Group's engagement, though admittedly indirect, with climate change since the early 1990s indicates the matters stressed in the report have been known to the Bank's actors and central to the Bank's agenda for this whole period. The unanswered question that runs through the report is why outcomes should be different now, and in years to come, than they have been in the past. As the report implies in chapter 7, box 7.1, what is needed most in the future elaboration of the entire IEG project is to

clarify and elaborate, in the light of its recorded behavior, the Bank's comparative advantage in the field of climate change.

Part A
There are very many discrete elements of the report that I found coherent, enlightening, and innovatively put forward. It makes a very useful contribution to the literature on energy and climate that would well be read within and outside the Bank Group. I'll list areas of treatment that, in my view, reinforce this conclusion.

1

The initial chapters on the relationships among energy growth, carbon emissions, and economic growth are concise and precise statements of what we know about these essential matters. They stress the critical points for the Bank Group and other major actors in the climate/energy intersection that poverty reduction and energy growth are not directly in conflict, that carbon and energy intensity are partially functions of natural endowments and partially products of clear choices about economic development paths, and that wide variation between nations in carbon emission performance is in part a function of energy policy and pricing. (Although given different labor, capital, and energy endowments, as well as the lack of understanding of carbon dynamics during the period in which basic patterns of economic development and resource use were set, the province and maintenance of these policies may themselves be subject to alternative interpretations.)

2

The tabular and analytical work on the carbon tax equivalence of recent increases in resource prices is original and quite helpful.

3

The case against subsidies and its political dynamics in the emerging era of high commodity prices and resource rent transfers summarizes well a mass of (fragmented) data clearly and deals nicely with the lack of basis for pushing these policies forward in the name of the poor, much better aided through other policy means.

4

The scale of the economic opportunities to reduce waste through energy efficiency and thereby avoid the construction of additional carbon-intensive generation is restated, but with apt attention directed to the gap between the technical and engineering potential of improving both economic and environmental performance and the far weaker experience of closing this gap. There are many particular and original observations throughout the report, based on case studies of the Bank Group's energy-efficiency program record (see #6 below) that contribute to the political economy or organizational theory explanations of why energy-efficiency gains are often ignored in practice.

5

The report is very informative in describing World Bank concentrations of loans and investments in specific dimensions of broad project categories. For example, in the area of energy efficiency, the bulk of projects and funds are placed in supply-side efficiency (equipment). Even in the limited set of projects aimed at managing demand- side efficiency (DSM), there is more emphasis given to technology (for example, CFL bulbs) than policy reforms (tariff decoupling—though it is shown that Bank Group electricity pricing reform should have a positive impact on the demand for energy-efficiency measures of all types). In the area of codes and standards, the emphasis is more on the elaboration and enactment of codes than on their monitoring or enforcement. Equally important, there are allusions to the role of organizational structures and incentives in producing these concentrations.

6

The report is replete with valuable and original observations that reflect the IEG author's substantial knowledge of the sectors and programs under review. They often stand in contrast to the lack of quality evaluation in other Bank Group processes designed to yield ongoing

increases in the productivity of investment. These observations most often are made in the course of case or project studies. Examples include:

a. DSM projects may often be undertaken as economical by utilities in developing countries that are forced by subsidized pricing to realize losses in some retail services.
b. In many cases there are serious questions about the causal impacts of Bank Group projects. Brazilian gains in conservation and energy efficiency in the 2001 drought period were more likely attributable to learning during mandatory rationing than codes or other policy reforms. Eastern European price reforms were more likely due to wide systemic movement toward markets than specific policy measures.
c. Even in cases where the economies of energy efficiency seem clear, subsidies to compact fluorescent lighting (ILUMEX) were not sustainable learning instruments that led to changed behavior when terminated.
d. The best energy-efficiency codes have little impact in the longer run without greater and sustained attention to monitoring and implementation capacity.
e. Favorable organizational image (public relations) was a more effective cause of reproducible behavior than other policies or subsidies in EGAT's (Thailand) success with compact fluorescent lightbulbs, indicating the potential of properly incentivized utilities.

7

The report details well how and why what appear to be win-win investments, especially in the area of energy efficiency, do not eventuate in a great number of instances. The roster of reasons varies from an absence of core collective goods like information to the presence of intranational resource transfer that requires either compensation or regulatory expropriation. But the report also makes it clear that many of these collective gains are efficient at the national level and that international transfers may be an unwise use of scarce financial resources. With these insights, it would seem that it would by now, after many years of Bank Group investment in this area, be standard operating practice within the Group to have developed effective analytical tools to discriminate between what should be done nationally and what internationally. However, there is no case made in the evaluation that any such tools have been consistently applied as normal use. The lack of attention over the years of Bank Group experience raises concerns about the incentives within the Group to manage these issues as well as might be hoped.

Part B

Before explaining my questions about the implications of the report for defining the comparative advantage of the World Bank Group in the area of climate change, I want to list a number of specific criticisms of the record made in the Report itself that are both persuasive and tempered.

1

Although there is increasing recent attention given to energy-efficiency support, especially by the IFC, when one considers the full spectrum of Bank Group investment in the energy/climate intersection (one in five projects has some connection to efficiency if a broader range of supply-side measures is considered), the relative proportion of the project funding going to energy efficiency has been less than optimal. Within this class of under-funded activities, the relative proportion to demand-side management is especially low in comparison to supply-side efficiency.

2

The report presents a good compilation of the mixed record of effectiveness of many of the core programs in the World Bank portfolio. These include the large number of investments in power sector reform, gas flaring in general and the Global Gas Flaring Reduction Partnership in particular, and energy pricing reforms. There are patterns observable in the variation in effectiveness within these programs. For example, fuel price reforms have been less successful than electricity price reforms; Eastern

Europe did better than large-scale fuel-producing nations. Moreover, the report notes very variable performance in project monitoring, analysis, and performance evaluation in the Bank's portfolio as well. (It is again surprising that there is as little systematic examination and learning from the variable record of performance as one would gather has occurred from a reading of the report's description of the materials to which it had access.)

3

There is good emphasis given in the report to the need for greater coordination across departments of the Bank Group to reduce intra-organizational stove-piping and the loss of potential benefits from a more comprehensive and systematic evaluation of the productivity of different investment options.

These three main themes form the logical and empirical basis for some of the key recommendations for reform. The first four recommendations are indisputable and well supported by the internal analysis of the report. These are: (1) focus on the removal of subsidies and provide targeted income compensation to the poor damaged thereby; (2) emphasize energy-efficiency opportunities and correct fuel and power prices to support these initiatives; (3) approach climate change systematically across the full range of World Bank country engagements because of the risk of perverse incentives under stove-piping; (4) improve the metrics and monitoring capacities to improve the information base on which such policy and program choices are made.

It is the fifth recommendation—that it would be better for the Bank to concentrate on those areas of the Bank Group's competitive advantage, namely, promoting policy and institutional reform—that I think would benefit from clearer and more explicit elaboration in future work. I do not suggest this because I disagree with the recommendation. I agree wholeheartedly that the weak record of positive results of all of our institutions around global climate change is generally best explained by hard problems associated with the implementation, monitoring, evaluation, and reform of misgovernance. What seems to merit further development in the light of this perception is more empirical evidence or organizational analysis that it is the comparative advantage of the Bank Group to be the agent best positioned to improve the record with regard to these agreed institutional objectives.

Just as the report correctly emphasizes that the problems with the realization in practice of win-win opportunities in theory lie often in political economy and organizational behavior, it may be useful in framing the future completion of this IEG project to ask directly why the Bank Group, after some 15 years of programming in the climate/energy intersection, continues to operate with a suboptimal investment portfolio and highly inconsistent analysis based on an inadequate information base. Project assessment has been narrow; carbon footprints have been haphazard; funding for renewables and energy efficiency has been generally low; implementation and monitoring are less attended than are normative prescriptions in policy-oriented activities. Are there systemic or institutional reasons that cause the persistence of these obvious and long-standing attributes of Bank Group practice? After initial experience with earlier programs that were subject to these same criticisms, why have there not been processes of systematic and sustained correction in later investment vintages? Would ongoing IEG work be more likely to induce positive change in the development in the Bank Group's program over time if there were more explicit discussion of the reasons that clarify why it has mainly stuck to a course that has long been subject to serious criticism?

We might here only speculate on types of organizational explanations that might be subjected to more intensive analysis to improve Bank Group practice by exposing the incentives that still are manifest in a relatively stagnant and problematic investment program. These might include arguments that an emphasis on normative economic prescription is too clear and too easy. This argument has been leveled at other dimensions of Bank Group programs by internal critics in areas including liberalization, privatization, and sectoral reforms. Related is the refrain that the path of transition from state-controlled to market-dominated economies was imagined as straightforward and technical, rather than profoundly political and conditioned by historical and institutional particularities in different countries. All of these claims could suggest the Bank Group has internal incentives to emphasize nonpolitical, often technical, remedies for poor growth performance; to stress upstream (technological) and normative solutions instead of downstream regulatory, behavioral, or implementation problems because the latter are relatively more constrained by fundamental concerns about intrusion into political operations that impose larger sovereignty conflicts.

An alternative line of explanation might begin in organizational sociology. The report notes that many of the relatively less frequent elements of Bank Group programs, like DSM or particular types of renewable generation, have been carried on under the particular aegis of GEF funding or are championed by small expert teams marginal to the larger Bank system. This observation suggests the foundational proposition of organization theory that large organizations have a core mission and an attendant adapted culture that dominates their priorities and performance. Such organizations respond to threats from the environment by establishing marginal groups that mediate external demands without disturbing core operations.

The Bank Group's core mission in this perspective is certainly to foster economic growth, with a strong amendment in the last decade to an express poverty alleviation orientation. This is reflected in an incentive system that concentrates on economic expansion and a commitment to short-run measures that bring poverty relief. Outcomes such as continued investment in energy infrastructure growth not necessarily constrained by environmental considerations (for example, coal plant investment) or technology diffusion rather than (longer-run) technology innovation would be expected in such an organizational culture explanation. (Conversely, focus on demand restriction might be less prized and reinforced because efficiency projects are complicated and staff-intensive, don't expend a lot of cash, and are less tangible and less prone to offer ceremonial occasions.)

These deeper issues of Bank organizational culture or internal incentives raise questions about what the report poses as the key issue going forward: what is the Bank Group's comparative advantage that should define its climate/energy strategy? With vast new resources coming onto the climate table, should primary responsibility be assigned to the Bank in allocating important segments of these resources, given its own institutional incentives? These questions may be premature in terms of the various phases of the complete IEG evaluation project. Major issues are not yet examined. These include both the contested record of the Bank Group in expending many times the funds on fossil fuel infrastructure financing than on noncarbon alternatives and the record of the Bank Group's carbon market initiatives. While the former is not addressed at all in the report, there are important anecdotal accounts of the latter.

Yet the preliminary work in the report also questions the Bank Group's early engagement with the CDM market in energy-efficiency financing, raising well-founded concerns about additionality if international funds are devoted to reducing costs of projects that are economically efficient at the national level. This is particularly true if continuing subsidies in retail prices reduce incentives for demand management. The report's chapter on gas flaring also analyses critically the Bank's use of CDM in cases where gas is not flared in the common cases where the regulated wholesale price of gas undercuts its collection and transmission, where electricity prices are held at levels too low to justify gas-fired generation, and where gas transportation projects that should be wholly economic at oil prices in excess of $40 per barrel do not take place because of risks of nonpayment from state-owned and run-off-takers. These prospective questions, yet to receive comprehensive IEG analysis, may be seen as challenges to the conclusory proposition that the Bank Group

should have a strong, though reformed, role in the growing world of carbon finance or climate policy.

In conclusion, at the end of discussing an excellent report, I wonder whether the report can best further the more effective resolution of such key climate change questions and help steer the Bank's internal evolution through more direct attention in the phases of the project to come to the issue of whether the Bank Group does have comparative advantages in climate in comparison to other potential climate institutions or to other public purposes the Bank Group might pursue.

Rajendra K. Pachauri

Chairman, Intergovernmental Panel on Climate Change; Director-General, Tata Energy Research Institute.

The report is comprehensive and reviews a range of World Bank activities that fit into an overall program related to climate change. Quite appropriately, the report traces the history and record of World Bank activities that are expected to have driven mitigation of GHG emissions over the years. The emphasis on institutional changes and reform measures is quite appropriate, because in the operations of the World Bank these assume logical primacy and should lead to outcomes in developing countries ensuring higher levels of energy efficiency and reduced emissions of GHGs as a consequence. It may be mentioned that the Intergovernmental Panel on Climate Change (IPCC) in its Fourth Assessment Report (AR4, 2007) has very clearly emphasized the importance of placing a price on carbon as perhaps the most effective policy measure for promoting technological change and other actions that could result in reduced emissions of GHGs. Hence, the viewpoint of the Bank on the issue of subsidies and their removal as well as rational pricing for different applications constitutes an important set of priorities that over a period of time can bring about change in the right direction. Addressing the assessment of several co-benefits, including lower levels of air pollution at the local level with attendant health benefits, higher security of energy supply, and the like in relation to mitigation of GHGs would have provided another dimension of externalities that should be part of economic decision making. This aspect has not been addressed adequately.

In my view, two additional aspects in preparing this report could have enhanced its value:

1. Research and development and technology issues for ensuring mitigation of greenhouse gases. While a number of technological innovations would generally flow from the developed to the developing countries, the need for customization of specific technologies to suit local conditions is an important aspect of technological change that perhaps deserved greater analysis and coverage in the report. This would also be justified by the fact that in several developing countries, technological capabilities have reached a level where they are making a significant difference in bringing about efficiency improvements and reduced emissions of GHGs.

2. The second subject on which greater coverage and targeted analysis would have been useful relates to adaptation to the impacts of climate change. It is very clear that effective climate policy in every country of the world would require a combination of mitigation as well as adaptation, most effectively to be conceptualized and implemented by the same organizations and authorities handling both. By not covering adaptation measures in adequate detail and confining the report essentially to mitigation, this dimension has been a loss in terms of the value of what is presented in the report.

All in all, this is a useful document, which, I am sure, will not only help the Bank in developing its own climate change portfolio in the coming years but would also be of value to policy makers and analysts in both the developing as well as the developed world.

Chapter 1

Evaluation Highlights

- The evaluation seeks lessons from policy experience in the energy sector to guide future policies on greenhouse gas mitigation.
- Mitigation of climate change will require a clean development path in both developed and developing countries.
- The central challenge of climate change mitigation is how to align national and global interests.

Person walks on a dirt road in Mali. Photo by Curt Carnemark, courtesy of the World Bank Photo Library.

Introduction, Scope, and Motivation

Climate and development are closely interlinked. Development has historically driven increased greenhouse gas (GHG) emissions. The buildup of these GHGs in the atmosphere is altering the global climate and threatening development.

The developed countries are responsible for most of the buildup of GHGs, and still emit far more per capita than the rest of world. But the developing and transition countries contribute the bulk of current emissions, and their contribution is swelling rapidly. To stabilize GHGs, all countries, both developed and developing, need to move toward a more sustainable growth path. To do so, however, developing countries will require financial and technological assistance. Appropriate policies will be critical for all countries.

This evaluation is the first of a series that seeks lessons from Bank experience on how to carve out a sustainable growth path. A premise of the series is that many climate-oriented policies and investments now under discussion have close analogues in the past. That is, policies and projects adopted with other aims—from fiscal discipline to biodiversity conservation—may have had significant impacts on GHG emissions or on adaptation to climate change.

A final, capstone summary to the evaluation series will offer a comprehensive look at the World Bank's role in climate change. This initial phase has a more limited scope. *First*, it introduces and sets the context for the series. *Second*, it tackles a small but ambitious segment of the climate development agenda as it pertains to the World Bank: key win-win policies related to mitigation. Figure 1.1 shows how this segment is nested within the broader issues. Table 1.1 describes how mitigation topics are divided between this phase and the next, whose project-level focus includes the International Finance Corporation (IFC) and the Multilateral Investment Guarantee Agency (MIGA).

This volume offers limited coverage of the Bank's role in climate change.

Although there are important overlaps, climate issues can be divided into those of adaptation and those of mitigation. Energy issues loom large in mitigation. (Emissions from deforestation, though large in the tropical world, have historically attracted less attention.) Within energy concerns, this volume focuses on World Bank–client engagement on policy interventions with the potential to confer immediate domestic benefits, while reducing emissions. These interventions have been pursued for many years and are still emphasized in discussions of current climate policy. Has the scope for such policies been exhausted? If not, what has been the record in pursuing them?

Figure 1.1: Intersection of Issues Related to Climate Change

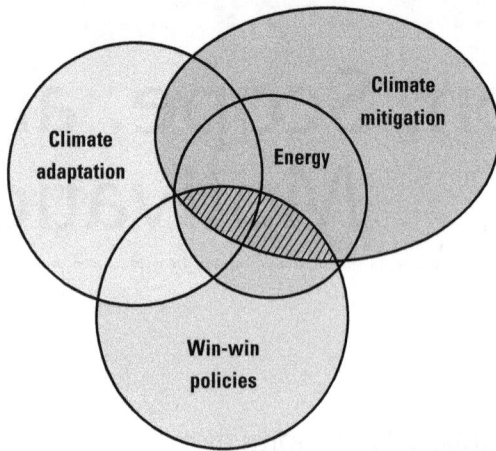

Compared with most Independent Evaluation Group (IEG) thematic studies, this volume places more emphasis on policy context. This is because the Bank lacks formal goals related to climate change against which evaluative assessments could be made. It also reflects a goal of drawing lessons for the Bank from external experience.

Energy policies are a significant concern for climate change mitigation.

The remainder of this section briefly sketches the striped territory shown in figure 1.1 and sets this evaluation and the rest of those in this series in context.

Confronting Inexorable Calamities and Unreckonable Risks

A changing climate threatens development and requires costly adaptations.[1] Higher temperatures bring inexorable calamities, with irreversible changes at specific locales. The sea will rise, exposing the large proportion of humanity that lives near a coast to inundation, flooding, and salinized water supplies. The Himalayan and Andean glaciers will melt, affecting water supplies to billions of people. Some areas could tip from semi-arid to arid, threatening the livelihood of some of the world's poorest people, and perhaps inducing mass migrations.

A changing climate threatens development.

Climate variability is increasingly unreckonable, complicating development planning.

Increased climate variability brings a host of risks. As temperatures rise more than 2° C over 1990–2000 levels, the frequency or intensity of extreme events such as hurricanes is likely to increase. Repeated weather shocks could threaten growth in poor, climate-vulnerable countries and regions. Risks are increasingly becoming *unreckonable*, which complicates planning for a wide range of endeavors. Because of climate change, the past is no longer a reliable guide to the future.

While climate models are improving, and show robust agreement about global trends, they often offer divergent forecasts of future average precipitation at the level of a specific province or river basin. Less predictable still are changes in the local likelihood of droughts, floods, and storms. Investments in water systems, agriculture, and disaster preparedness thus have to hedge bets against an increasingly uncertain future, an expensive undertaking. At the global scale, there is a small but growing chance of a planetary catastrophe—an increase of 5° C or more that would lead to profound and universal changes in sea level, weather, and ecosystems (Stern 2007).

Adaptation to these changes has to be combined with mitigation of their severity. Indeed, in the short run there is no way to alter the climate changes that are already in train, so that adaptation is essential. The longer the horizon, however, the more leverage there is to moderate GHG emissions and reduce the worst long-term risks.

The United Nations Framework Convention on Climate Change (UNFCCC) requires that the atmospheric concentration of GHGs—now at 430 ppm CO_2e (parts per million of carbon dioxide equivalent)—be stabilized at safe levels. "Safe" levels are debated: the *Stern Review* advises a target between 450 and 550 ppm to minimize the chance of catastrophic outcomes; others, worried about crossing a tipping point to accelerated CO_2 release, recommend lower levels. Global models (IPCC 2007a) show that to stabilize CO_2e concentrations below 535 ppm, global emissions must begin to decrease before 2020—sooner, if more stringent limits are sought.

Table 1.1: Topical Map of Issues in the Climate Evaluation Series
(Topics in shaded areas are covered in this phase; those in unshaded areas will be discussed in the second phase of the evaluation.)

Issue	Policies: design and implementation (IDA/IBRD)	Investments in technologies, facilities, hardware, financial intermediaries (IDA/IBRD/IFC/MIGA/carbon finance)
Energy pricing	National adoption of policies that remove energy subsidies or rationalize energy prices	Impact of power pricing policies on specific investments in renewable energy and energy efficiency
Energy efficiency	Policies (in addition to pricing) that encourage energy efficiency, with emphasis on end-user and demand-side efficiency	Efficiency finance, including ESCOs; facility-level investments in demand- and supply-side efficiency
Gas flaring	Natural gas pricing policies and gas flaring reduction; GGFR experience	Not covered
Transport	Fuel pricing policies	Transport projects
Renewable energy	Renewable energy policies (feed in tariffs) affecting investments	Investments in specific technologies (wind, water, and the like)
Reduced emissions from deforestation and forest degradation	Not covered	Protected areas, enforcement of anti-deforestation laws, community forests

Note: IDA = International Development Association; IBRD = International Bank for Reconstruction and Development (World Bank); GGFR = Global Gas Flaring Reduction Partnership; ESCO = energy service company.

Developed countries are largely responsible for the current level of climate change, and emit far more GHGs per person than the developing countries. Climate stabilization requires essentially a complete phase-out of these emissions in the long run, with significant near-term progress toward that goal. The UNFCCC calls on developed countries to take the lead in mitigating emissions.

However, climate stabilization is not possible without the availability of a clean development path in the developing and transition countries. Even complete elimination of developed-country emissions would not suffice by itself. The Bali Action Plan (UNFCCC 2007) commits all members of the UNFCCC to the pursuit of "deep cuts in global emissions," "in accordance . . . with the [UNFCCC] principle of common but differentiated responsibilities and respective capabilities, and taking into account social and economic conditions and other relevant factors." That means

finding a better path to wealth for the developing countries than that trod by the developed countries. Both the Bali Action Plan and the UNFCCC call for developed countries to provide "new and additional" funds and technology that would allow the developing countries to do this.

Adaptation must be combined with mitigation.

Near-term actions can shape that long-term trajectory, with big consequences for long-term growth and emissions. The concern is with lock-in. For example, poorly insulated buildings and inefficient coal plants built today will be in place for decades, consuming money and emitting CO_2. Energy subsidies not only stimulate inefficient, emissive energy use; they also generate strong constituencies for those inefficiencies, which makes them difficult to reverse. Similarly, it is easier to fight congestion and pollution by establishing

Stabilization of climate change requires a clean development path in both developed and developing countries, but developing countries need financing.

Actions taken now can shape long-term emission patterns.

road-user charges before car ownership is widespread, than after.

Three Approaches to Greenhouse Gas Mitigation

Mitigation of GHGs presents a classic problem in environmental economics. A country that reduces its emissions typically incurs costs, but reaps only a small proportion of the global benefits of an improved climate. So countries are not motivated, individually, to undertake the optimal degree of global mitigation. IEG's *Annual Review of Development Effectiveness 2008* discusses the challenge of global public goods at length (IEG 2008a).

One approach is to seek win-win policies and projects, but implementation is often impeded by regulatory barriers, coordination problems, vested interests, and institutional and market failures.

There are three prominent policy approaches to this dilemma. They broadly represent the World Bank's past approach to climate change and are consistent with the UNFCCC principle of "common but differentiated responsibilities" of developed and developing countries. The first is to seek win-win (or no regrets) policies and projects. These not only provide global benefits in reducing GHGs but also pay for themselves in purely domestic side benefits such as reduced fuel expenditure or improved air quality. (See upper-right quadrant of figure 1.2.) For instance, countries could remove fossil fuel subsidies, thereby curbing GHGs, improving local air quality, and freeing government funds for better-targeted social programs.

If win-win policies were easy to implement, they would have been put in place long ago. But regulatory barriers, coordination problems, institutional failures, opposition by vested interests, or market failures impede them. That is, the nation may benefit as a whole, but there are groups who lose under win-win policies. External finance, such as development lending or concessional funds, could be used to facilitate adjustment to the win-win policies.

Another approach is to seek compensation from the global community for countries that provide GHG reductions.

A major goal of this evaluation is to provide insight into the potential scope for win-win policies and

into strategies for designing and implementing them in the face of various barriers. Some analysts (IEA 2006; McKinsey Global Institute 2008) see tremendous untapped opportunities for win-win policies and projects; others are skeptical. There are questions about both the applicability and the feasibility of implementing these policies. The World Bank Group's extensive involvement in supporting win-win climate policies has sometimes been framed in climate terms, but more often justified on purely domestic, sectoral grounds.

The second approach is to seek compensation from the global community for countries that provide GHG reductions. This approach is attractive to a country if the combination of compensation and domestic side benefits outweighs the costs of policy adoption. (See carbon finance segments in figure 1.2.) It underlies the UNFCCC principle of "common but differentiated responsibilities." This principle reflects the unwillingness of developing countries to accept limits on emissions or incur costs to limit emissions. They point to much higher per capita emissions by developed countries, and have called on them to take the lead in global reductions.

Compensation could take the form of grants to cover the additional costs of providing reductions (an approach that has been used by the Global Environment Facility [GEF]) or payments for the reductions themselves (the carbon market approach). For convenience, this report will refer to both as carbon finance. The Clean Development Mechanism (CDM)—a creation of the Kyoto Protocol—is the biggest vehicle for carbon finance, which allows developed countries to meet their climate obligations by paying for emissions reductions in the developing world. The CDM is currently restricted to project finance and excludes support for GHG-reducing policy reforms. The Kyoto Protocol also sets up incentives for some developed countries to fund reductions in transition economies. GEF projects can also be viewed as a kind of carbon finance, though funds are usually represented as supporting catalytic actions rather than as compensation. The World

Figure 1. 2: Global and Domestic Benefits

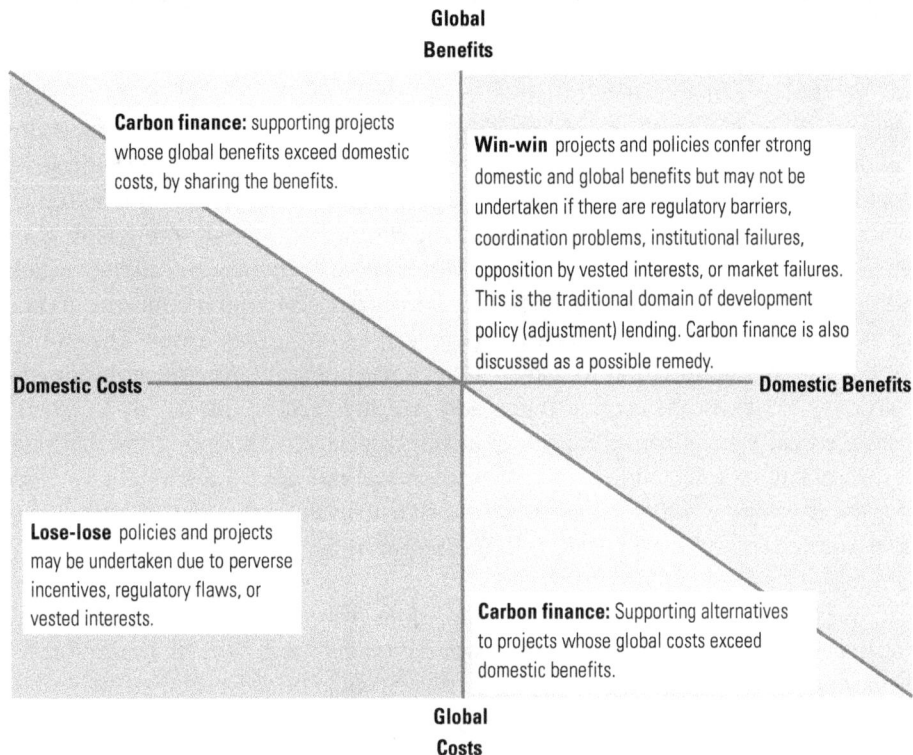

Global
Benefits

Carbon finance: supporting projects whose global benefits exceed domestic costs, by sharing the benefits.

Win-win projects and policies confer strong domestic and global benefits but may not be undertaken if there are regulatory barriers, coordination problems, institutional failures, opposition by vested interests, or market failures. This is the traditional domain of development policy (adjustment) lending. Carbon finance is also discussed as a possible remedy.

Domestic Costs

Domestic Benefits

Lose-lose policies and projects may be undertaken due to perverse incentives, regulatory flaws, or vested interests.

Carbon finance: Supporting alternatives to projects whose global costs exceed domestic benefits.

Global
Costs

Bank has been extensively involved in developing and implementing CDM and GEF projects.

A third, hybrid approach promotes research and development in clean technologies. Immature technologies are expensive and risky, so few people will use them without incentives. For instance, solar power is cleaner but more expensive than fossil fuels for grid-connected electricity. But the cost of solar power, like most technologies, decreases as there is more and more experience with manufacturing and using it. For this reason, industrial strategists advocate pushing technologies down the learning curve, so that they end up in the win-win segment. GEF's Operational Program 7 has attempted to do this with concentrated solar power and other technologies.

Priority Areas for Evaluation Related to Mitigation

There is an immense range of activities, across many sectors, that can contribute to climate change mitigation. To focus this evaluation series, the following criteria were considered:

- Large potential for mitigation at low cost in developing and transition countries
- An evaluable World Bank Group record, including incorporation in policy and strategy statements
- Relevance to future World Bank Group strategy
- Solid scientific basis for linking activities to GHG emissions.

With respect to mitigation potential, IPCC (2007a, p. 632) presents a synthesis of current estimates, using reduction potential relative to a business-as-usual baseline in 2030 as a benchmark. For the developing world (that is, outside the Organisation for Economic Co-operation and Development [OECD] and economies in transition), it estimates that there are 2.7 billion tons of CO_2e of negative-cost potential savings through end-use

A third approach is to promote research and development in clean technologies.

7

efficiency in commercial and residential buildings, including appliances. The availability of negative-cost opportunities indicates market failures in need of policy attention. This compares with 0.1 billion tons in negative-cost transport opportunities. In power generation, IPCC estimates available savings of 0.8 billion tons for developing countries at a cost of less than $20 per ton (possibly including some negative-cost options) from cleaner fuels, renewable energy, and increased generation efficiency; another 1.25 become available at costs up to $100 per ton. End-use efficiency in industry offers 0.6 billion tons at less than $20 per ton. Agriculture and forestry account for about 1.1 billion tons each at that cost. Low-cost (less than $20 per ton) reduction opportunities, across all sectors, amount to 6.9 billion tons for the developing world, 1.2 for the economies in transition, and 4.5 for the OECD; 1 billion tons are regionally unallocated.

This overview suggests that policies affecting end-user energy efficiency stand out as the area with the single greatest potential for emissions reduction, and at potentially negative rather than positive cost—a win-win option. As subsequent chapters of this evaluation will show, it is an area that the World Bank has stressed in sectoral strategies, and where it has deployed project, analytic, and capacity-building effort. One set of win-win policies—removal of energy subsidies—potentially promotes not only end-user efficiency, but also supply efficiency and renewable energy. Here, too, there has been extensive World Bank involvement. So energy pricing policies, and non-price-related energy-efficiency policies, constitute one focus of IEG's evaluation series.

Policies affecting end-user energy efficiency stand out as the area with the single greatest potential for emissions reduction.

All models of global mitigation show that exploitation of win-win opportunities is insufficient to stabilize GHGs in the atmosphere. Massive investments in low-carbon energy technologies—the menu includes solar, wind, hydropower, nuclear, and carbon capture and storage—will be necessary. Much of this investment will take place in the developing

This report is mainly concerned with win-win policies in the energy sector.

world, where energy demand is growing rapidly. The complexity of the international negotiations around climate change revolves largely around how the burden of abatement costs—the incremental costs of GHG-reducing technologies—will be shared. Under the Kyoto Protocol, developed countries take on obligations for reducing emissions but can satisfy these obligations, in part, by financing emission reductions in the developing world. The Bali Action Plan calls for provision of new and additional financial resources for developing countries to address both adaptation and mitigation. The World Bank Group has been involved in mobilizing public and private sector funds to support the incremental costs of adopting and diffusing low-carbon technologies. So this, too, is a focus of the climate evaluation series, though not of the current volume.

Emissions from deforestation in the developing world are significant. Reduction of deforestation, in theory, could be accomplished at low cost and would offer numerous local side benefits (Chomitz and others 2007). The World Bank has been a supporter of forest conservation and sustainable use; lessons from that experience are relevant to plans to use carbon finance to support reduced emissions from deforestation and degradation (REDD). In contrast, while agriculture is known to be a significant source of GHG emissions, and there are prospects of win-win approaches, there is much less of an evaluable record to examine. Some of the basic science is still imperfectly understood, and measurement of emissions from nonpoint sources (livestock, rice fields) is difficult.

Scope and Methods of This Evaluation

Table 1.2 places this volume within IEG's examination of climate issues. This evaluation is concerned with the first of the three mitigation approaches—the win-win policies. It confines its attention to the energy sector, where experience is greater and where there has been more attention to climate implications. Because of the policy focus, it is mostly restricted to the experience of the World Bank (International Bank for Reconstruction and Development [IBRD] and the International Development Association [IDA]), although IFC experience is referenced where useful for context and comparison. An ongoing IEG evaluation is examining the Bank's recent implementation of its forest policy.

Phase II of the climate evaluation will look at the second and third approaches to mitigation. Drawing on and expanding an earlier IEG report on renewable energy (IEG 2006b), it will review the World Bank Group's record in promoting investments in renewable energy and energy efficiency. The World Bank Group has used different units and financing mechanisms— including carbon finance, IFC investments, GEF grants, and IDA lending—to promote technology diffusion or to compensate countries for the cost of adopting technologies with global benefits. This phase of the evaluation will also assess the institutional contributions of the Bank's Carbon Finance Unit in spurring global transfers and aspects of the Bank's forest experience relevant to the REDD agenda. Table 1.1 shows the division of mitigation topics between the two phases.

A planned third phase will look at emerging practice in adapting to climate change. IEG has also undertaken or planned a number of

thematic studies that are relevant to climate change adaptation. These include published evaluations of the power sector (IEG 2003), of renewable energy (IEG 2006b), and of natural disaster prevention and relief (IEG 2006a). Ongoing evaluations of World Bank support for water management and for agriculture provide background for adaptation issues.

The plan for this evaluation is as follows. Chapter 2 uses cross-national data to illustrate the link between development and energy-based emissions, including the scope for policies to weaken this link. It presents a general framework for understanding energy policy-to-emissions links, which are numerous and complex. It also examines the interlinkage between the energy access and climate mitigation agendas.

Later phases of the evaluation will examine the other approaches to mitigation.

Chapter 3 is a selective review of World Bank involvement in issues related to climate change mitigation. It traces the treatment of climate change in sector strategic documents over the past 15 years. It gauges the extent and correlates of attention to climate and related issues in the country strategies of the largest emitters among the Bank's clients.

Table 1.2: IEG Evaluations Relevant to Climate Change

Theme	Coverage	Evaluation	Date
Climate mitigation	National policies, concentrating on energy	Climate Change, Phase I	2008
	Forest policies and projects	Evaluation of Bank's Forest Strategy	2009
	Low-carbon investment projects, technology diffusion, carbon finance	Climate Change, Phase II	2009
Climate adaptation	Project and policy experience specifically related to adaptation	Climate Change, Phase III	2010
Capstone summary of climate evaluation	Synthesis of Phases I–III		2010
Sectoral evaluations on related topics		Water Sector	2010
		Agriculture	2009
		Renewable Energy	2006
		Natural Disasters	2006
		Power Sector Reform	2003

While the evaluation focuses on learning lessons for GHG reduction, it is important to acknowledge that the Bank Group's activities can potentially promote GHG emissions as well as mitigate them. While it is beyond the scope of the evaluation to assess the Bank Group's carbon footprint, chapter 3 reviews precedents and approaches to doing so, including the use of carbon shadow pricing in project appraisal and portfolio decisions.

Chapters 4 and 5 examine two related areas that have large economic and environmental scale and are thought to offer large win-win opportunities: energy pricing and subsidies and energy efficiency. In both areas, literature reviews establish the scope for economic gains and for emissions reductions. Special attention is paid to compilation of evidence on the impact of price reform on poor people.

For both areas, content review of Bank lending over 1996–2007 (with selective attention to earlier years) identifies policy components of development and investment lending, again permitting assessment of patterns and correlates of engagement. Documentary and statistical evidence and interviews were used to assess patterns and correlates of engagement and of outcomes. Engagement and outcomes on pricing were assessed in more depth in the countries with the largest absolute levels of subsidy. Because of the complexities of attribution and of modeling, it was not, in general, possible to make quantitative estimates of the impacts on GHG emissions.

Chapter 6 is a case study of an apparently win-win topic: gas flaring. The topic is interesting because of its magnitude (more than 400 million tons of CO_2e per year), the links to policy and to carbon finance, and the existence of a World Bank–led initiative for flaring reduction.

A final chapter summarizes findings and synthesizes cross-cutting recommendations. It also looks forward to the second phase, presenting an analytic framework for thinking about clean technology diffusion.

This volume does not offer a comprehensive assessment of the World Bank's role in climate change. It leaves out many important areas of engagement. It does not discuss forest issues, and contains only a superficial discussion of policies related to renewable energy. It does not cover the Bank's advisory and capacity-building efforts related to the Kyoto Protocol.

The forthcoming second phase, with its concentration on the project-level experience with low-carbon technologies—including renewable energy and energy efficiency—will cover much World Bank Group activity explicitly oriented to mitigating climate change, including the role of the carbon funds.

Chapter 2

Evaluation Highlights

- Emissions levels are closely tied to income level and population, but policy has substantial leeway to reduce emissions.
- Fuel subsidies increase emissions.
- Poor countries emit relatively small amounts of GHGs, and the benefits of increased electricity access far outweigh the costs.

Indonesian motorists line up for gasoline in Bogor. Photo ©Dadang Tri/Reuters/Corbis, reproduced by permission.

National Policies and Climate Change

This chapter looks at the relationship between development and energy-related GHG emissions. It examines the degree to which the Bank's support for clients' growth and poverty reduction places pressure on GHG emissions, with particular attention to the issue of energy access for the poorest. It also assesses the scope for policies and investments to affect national GHG emissions.

Energy, CO_2, and Development: A Strong but Pliable Relationship

Energy use is a large and growing source of GHG emissions. In transition economies, combustion of fossil fuels (including transport and industry) accounts for almost 90 percent of emissions. In developing countries, 43 percent of emissions are from energy and industrial processes, 37 percent from deforestation and land use change, and 16 percent from agriculture.[1] The energy proportion will rise over time, since energy use is growing faster than emissions from deforestation.

Emissions rise with income and population, and are higher in colder climates. CO_2 emissions are deeply connected, through energy use, to development. Figure 2.1 shows the strong relationship between per capita income and per capita emissions of energy-related CO_2. This relationship is tighter than the more frequently displayed relationship between CO_2 and gross domestic product (GDP), because countries shift into and then out of manufacturing as income increases. It underlines the expectation that

development will generally result in higher emissions. It is crucial to keep in mind that the graph is logarithmic: emissions per capita of low-income countries are only a small fraction of those of high-income countries. There is a 600-fold difference in per capita emissions between the highest- and lowest-emitting countries shown.

Emission intensity is linked to per capita income.

Figure 2.1 also distinguishes among countries with different climates, indexed by heating degree days. The warmest countries are represented by triangles, temperate countries by circles, and the coldest countries by pluses. Colder countries tend to be wealthier, but the relationship between income and emissions is less pronounced in this group. Income and heating need together explain more than 85 percent of the variation in per capita emissions.

But some countries are much less intensive than others at the same income level.

Nonetheless, countries vary significantly in their emissions intensity, even after adjusting for level of development. Although the relationship in figure

13

Figure 2.1: Per Capita Energy Emissions and Income, 2004

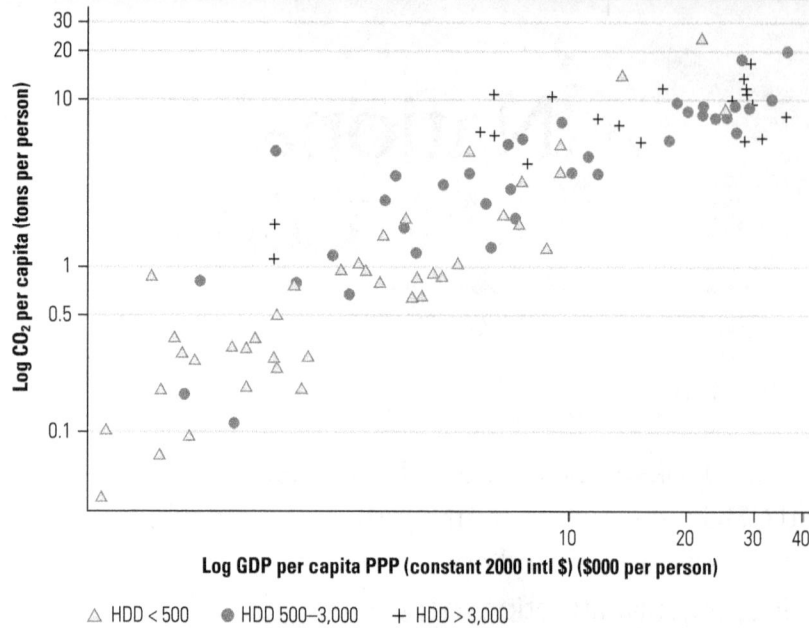

△ HDD < 500   ● HDD 500–3,000   + HDD > 3,000

Source: IEG calculations based on International Energy Agency and World Resources Institute data.

Note: HDD = Heating degree days. Countries with population < 4mln (2004) excluded: PPP = purchasing power parity.

2.1 is strong, it is a thick band, not a thin line. Most countries lie along the center of the band, but, holding income constant, there can be a sevenfold difference in emissions intensity. In other words, some countries emit much less than peers at similar levels of development, and some emit much more.

This variability reflects some leeway in the linkages between GDP and energy use, between energy use and fossil fuel consumption, or between fuel combustion and CO_2 emissions. The energy-GDP ratio depends on a nation's mix of agriculture, manufacturing, and services—more energy is required to produce a dollar's worth of aluminum than an equivalent value of cassava or insurance policies. It also depends on how efficiently firms and households use that energy—for instance, on how well they insulate their homes and factories. The emissions-energy ratio depends not only on the role of fossil fuels versus renewables, but also on the precise mix of fossil fuels and the technologies used to burn them (box 2.1).

What determines a country's emissions level relative to its peers? Chance, to some extent—the luck of being endowed with coal or oil deposits. The use of hydropower is a significant determinant of emissions intensity, and reflects both water resources and energy policy.[2] Specialization in fossil fuel-intensive exports (such as refinery products, steel, and aluminum) will boost relative emissions, especially for small or poor countries. Measurement error also plays a role, since it is difficult to measure CO_2 emissions comprehensively. However, the relative emissions may in part reflect policy decisions—on energy pricing, for example. So, while this report attaches neither blame nor praise to relative emissions, it uses them as a diagnostic, a useful but imperfect indicator of the scope for reducing GHGs at a given income level.

Some countries have moderated their emissions per capita despite increased income. Figure 2.2 shows how absolute levels of per capita income and emissions changed over the period 1992–2004 for all countries. Most countries have moved up along the diagonal, increasing both income per capita and emissions per capita. But a few countries (blue arrows) have moved down and to the right, increas-

Box 2.1: Emissions Intensities of Power Supply

Generators transform energy into electricity. The emissions intensity of supply—CO_2 emissions per kilowatt-hour (kWh)—depends on the source of primary energy and the efficiency with which that energy is transformed into electricity.

Nonfossil energy sources—wind, solar, nuclear, sustainably grown biomass, and some kinds of water power—can produce power without net CO_2 emissions (setting aside the CO_2 emitted in the course of manufacturing turbines and other equipment). An Intergovernmental Panel on Climate Change (IPCC 2007b) review found that most hydropower plants offered "low net GHG emissions," but that scientific uncertainties remain. In the tropics, emissions of methane—a more potent GHG than CO_2—from shallow plateau-type reservoirs and from reservoirs with low power-to-flooded-area ratios have been found to be relatively large, but are smaller from deep reservoirs. Emissions are thought to be low from most boreal and temperate reservoirs (UNESCO 2006) and are not an issue for run-of-river plants that have no reservoir.

As a rule, gas generates less CO_2 per unit of heat than oil, and oil generates less than coal. Fuel switching is thus an important strategy for emissions reduction. Even for a specific fuel there are big variations in power plant efficiency—the proportion of energy that gets transformed into electricity or commercially valuable heat. Small plants tend to be less efficient in producing heat than larger ones, and hence more CO_2-intensive. Cogeneration—the combined production of heat and power from a single plant—saves energy and emissions compared with separate production of these two services. In principle, power plants can reduce their emissions to zero by capturing and burying CO_2 emissions from their smokestack, but carbon capture and storage technologies are still experimental.

The table below illustrates the range of emissions intensities associated with different fuels and technologies based on new plants. Life-cycle measures are higher. Liquefied natural gas (LNG) requires substantial energy for liquefaction and transport, but on a life-cycle basis, a modern LNG-fueled generating plant is still 38–47 percent less carbon-intensive than a modern coal plant (Hondo 2005). A substantial amount of electricity can be physically dissipated (as opposed to stolen) in transmission and distribution. These losses would have to be taken into account to estimate emissions per kWh consumed by end-users.

CO_2 Emissions of New Power Plants by Fuel and Technology (grams per net kWh)

Source	NETL	ESMAP
Gasoline 1 kW		1,500–1,900
Coal subcritical	855	880
Coal supercritical	804	
Coal IGCC	752–796	700–750
Diesel 5 MW		650
Oil combustion turbine		780
Oil combined cycle		520
Gas combustion turbine		600
Gas combined cycle	362	400

Sources: ESMAP 2007; NETL 2007.
Note: ESMAP = Energy Sector Management Assistance Program; NETL = National Energy Technology Laboratory.

ing per-person income while decreasing per-person emissions. Many of these are transition economies that also managed to drastically decrease emissions per dollar of GDP. Most of these countries began this reduction from a very high relative level of emissions through substantial restructuring of their economies and adjustments in energy prices. In addition, some developing countries—such as Botswana, China, *Emissions levels are related to natural resource endowments, but policy decisions also play a role.*

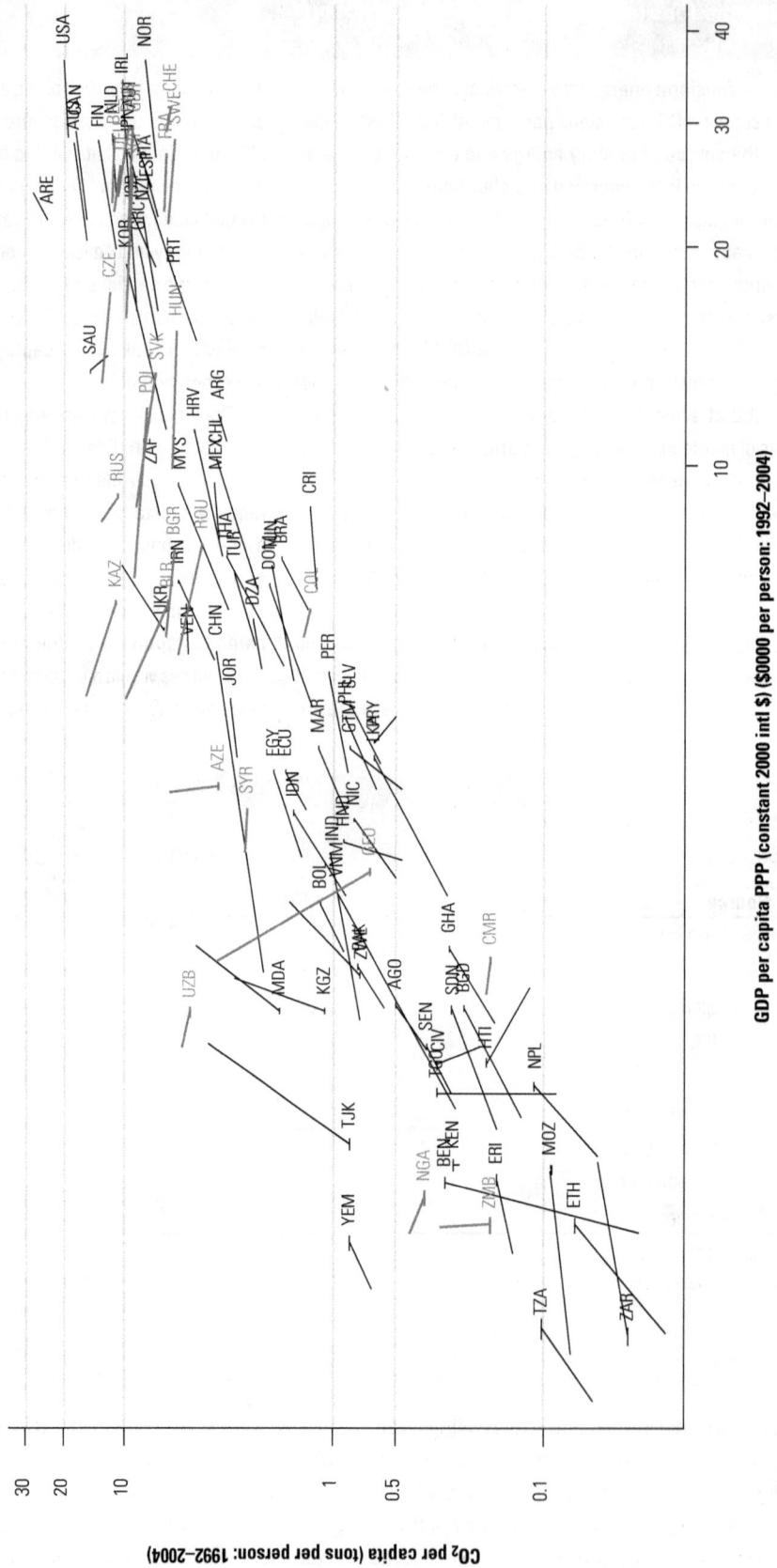

Figure 2.2: Absolute Changes in Emissions and Income, 1992–2004

CO2 per capita (tons per person: 1992-2004)

GDP per capita PPP (constant 2000 intl $) ($0000 per person: 1992–2004)

Source: Authors' calculation based on International Energy Agency data..

Note: Blue = CO₂ 1992 > CO₂ 2004 and GDP 1992 < GDP 2004; countries with population < 4 mln (2004) excluded.

and India—registered large gains in income per capita with relatively modest gains in emissions.

In sum, the tide of development strongly pulls countries to higher emissions per capita. But some countries swim across this current. Policy—at least potentially—has substantial leeway to reduce emissions. Next we look at the pathways through which this might occur.

Policies and Institutions Can Make a Big Difference

Supply, transformation, and demand policies affect the scale and mix of energy use. Table 2.1 presents a policy typology that guides this evaluation of energy policies and emissions. At the center of the diagram is infrastructure for power generation and transmission. Emissions go up with the scale of generation: the total amount of power produced.

Emissions also depend on the mix of primary energy used to generate electricity and on how efficiently power is generated and transmitted. Coal combustion releases about a ton of CO_2 for each megawatt-hour produced (with considerable variation, depending on plant efficiency); natural gas releases about half as much; wind and run-of-river hydropower release none. So, from an emissions perspective, it matters a great deal whether a country builds coal, gas, or hydroelectric power plants; whether its fossil fuel plants squeeze more or less electricity out of each ton of carbon burned; whether low-emissions plants are dispatched in preference to higher-emissions

ones; and how much energy is lost in transmission before it reaches homes and factories.

Scale and mix of power generation are shaped by three related sets of policies: those affecting supply of primary energy, power plant technology choice, and demand. On the supply side, pricing and regulatory policies affect the relative price and availability of coal, oil, gas, and hydro. Energy availability is an obvious determinant of power system technology. But power sector regulations matter too, and can affect the efficiency of power transmission and distribution—an important but sometimes overlooked factor affecting emissions. On the demand side, price policies and efficiency policies guide people, companies, and government agencies as they choose how much electricity and heat to consume.

Public policies also shape the scale and mix of energy use for transport. Supply-side policies include those on investments in roads and transit and public transport systems. Demand-side policies include fuel prices, vehicle taxes and standards, and urban planning.

Table 2.2 sketches specific pathways through which broad policy reforms can affect emissions intensity at the provincial or national level. Note that these pathways can affect emissions directly by influencing demand for energy, the source of energy, or the efficiency with which energy is used. They can also affect emissions indirectly by

Emissions depend on the mix of energy used to generate electricity and the efficiency of generation and transmission.

Table 2.1: How Policies Affect Energy-Related Emissions

Supply	Transformation	Demand
• Primary fuel price and availability	• Renewable portfolio standards; pollution regulations	• End-user tariffs and collections: electricity, heat, gasoline, diesel
• Coal regulation and mining subsidies		• Demand-side management
• Gas regulation, pricing, and infrastructure		• Building, appliance, vehicle standards and regulations
• Flaring regulation		• Public procurement

Source: Author.

Table 2.2: Pathways from Policies to Emissions

Policy	CO_2 impact (+ indicates an increase in CO_2 intensity)
Policies affecting supply	
Remove subsidies to or protection of coal supply or transport; shut down uneconomical coal mines	−
Remove price controls on natural gas supply	−
Remove regulatory barriers to use of associated gas from oil fields	−
Provide capital subsidies to generation (from domestic or international sources)	+
Coordinate international energy infrastructure	−
Privatize generation	±
Regulate hydropower facilities	±
Incorporate energy security considerations into energy sector expansion plans	±
Promote renewable fuels for power generation	−
Regulate and enhance enforcement of limits on industrial pollution and pollution from power generation	−
Promote bus rapid transit	−
Shift buses to compressed natural gas	−?
Policies affecting demand	
Remove subsidies or price caps on electricity; increase collection rate of fees	−
Institute time-of-use charges for electricity	−?
Remove consumer subsidies for heat while enabling control of heat use	−
Remove subsidies for kerosene, gasoline, and diesel fuel	−
Implement efficiency standards for buildings and appliances	−
Promote financing for energy efficiency	−
Promote more efficient urban land use	−

Policies regarding energy supply, technology choices, and demand affect the scale and mix of power generation.

stimulating or stunting growth, given the close relationship between income and emissions. (In some cases, these policies will reduce emissions intensity, but increase power production, so it is possible that absolute emissions could increase.)

As an illustration of the link between policy and emissions, consider the relationship between diesel pricing and relative emissions. Diesel is a globally traded commodity, but tax and subsidy policies cause its price to vary widely among countries. (Diesel is more likely than gasoline to be subsidized.) Unlike most other energy prices, retail diesel prices are readily observable. The GTZ

Fuel subsidies are associated with increased emissions.

(German Technical Cooperation) (GTZ 2007) regularly collects this information and suggests that the price relative to the U.S. price can be viewed as an indicator of subsidies or taxes, since the U.S. price is close to a free market value.

Figure 2.3 shows this relationship for 2004. There is a relatively strong negative correlation ($\rho = -0.39$) between diesel price and relative emissions. Note the well-known tendency for oil producers to subsidize fuel. Most striking, essentially all countries that maintain diesel prices below half the reference level exhibit high relative emissions—on average, 91 percent above their peers. This differential is too large to be understood merely as the effect of excessive diesel use (though that may be part of the story).

Pathway
Shift to lower-carbon energy sources
Increases supply of gas, induces shift away from coal or oil
Increases supply of gas, induces shift away from coal or oil; harnesses energy otherwise wasted in flaring
Favors generation over end-use efficiency
International sharing of hydropower supply and coordination on natural gas pipeline can substitute for smaller-scale, less-efficient coal or diesel power generation
Probably favors gas-based generation over coal (reducing emissions) or hydro (increasing emissions); should promote generation efficiency
Depending on their application, environmental and social regulations could expand or contract the development of hydropower facilities and restrict or allow plants that create methane emissions
Could promote a shift toward coal (if reserves are available) or renewables, boosting or decreasing emissions intensity
Substitutes for fossil fuels
Particulates and sulfur oxide (SO_x) pollution from power generation and industrial activity can be mitigated in part through greater efficiency in the combustion of coal and oil, through cogeneration, and by switching to gas and renewables. However, there are also pollution mitigation options that do not involve GHG reductions.
Reduces fuel consumption by shifting passengers from cars and reducing congestion
Possible reduction in emissions intensity
Where consumers have unrationed access to subsidized energy, higher prices will lead to reduced consumption and emissions. Where low prices have led to inadequate investment, removal of subsidies could result in expanded supply of grid-based power and decreased use of small or captive plants, probably with a net decline in emissions intensity.
Depends on fuel source for peak versus base load; could reduce emissions where peak load is met with old, inefficient generators
In many transition economies, heat has been subsidized, and consumers lack both the ability and incentive to economize on heat use.
Higher prices will lead to reduced consumption and emissions.
Reduces energy consumption (unless demand for energy is extremely price-elastic)
Reduces industrial and commercial energy consumption
Reduces fuel consumption by reducing the demand for transport

Rather, the diesel subsidies may reflect more pervasive energy-price distortions.[3]

GHG Mitigation Need Not Compromise the Pursuit of Energy Access for the Poorest

About 2 billion people lack access to electricity. Electricity provides poor people with a broad range of social and productive benefits and is widely viewed as an important tool for achieving the Millennium Development Goals. Does the goal of mitigating GHGs stand in the way?

It need not. Figure 2.1 shows that poor countries emit only a tiny fraction of the per capita emissions of rich ones. A rough calculation shows that providing 2 billion people with basic electricity access—one kWh per household each day—would boost world GHG emissions by less than 0.4 percent, even if power were provided entirely by the most carbon-intensive means. The rest of the world increases its carbon emissions by this much about every two months.

Poor countries emit only a tiny fraction of the per capita emissions of rich countries.

The benefits of electricity access to the poor also far exceed any conceivable damages from the associated emissions. An IEG review of willingness-to-pay for grid-connected electricity found values of $0.47 to $1.11 per kilowatt-hour[4] (IEG 2008c). A project to meet unserved needs through diesel power (a typical option in Africa) would result in emissions of 600–1,000 grams per kWh. So even if damages were assumed to fall

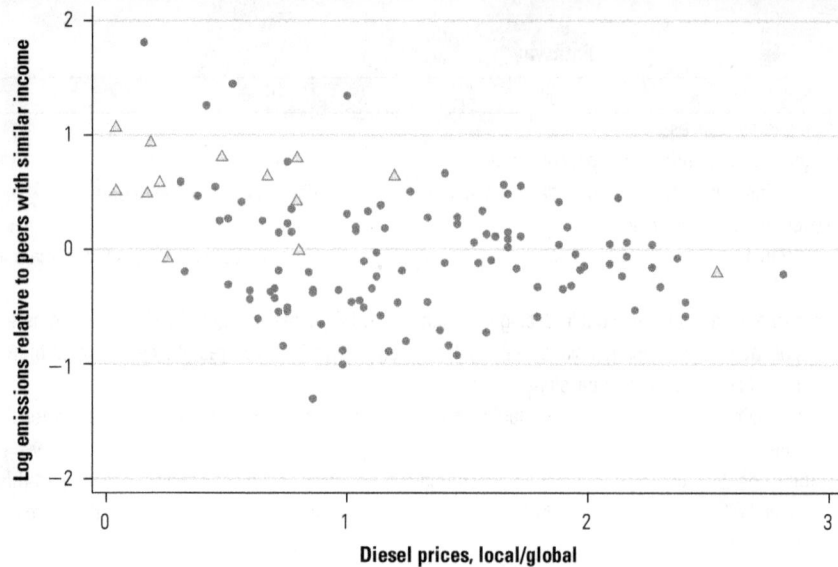

Figure 2.3: Relative Emissions Are Higher in Countries with Diesel Subsidies

Sources: Relative emissions: Chomitz and Meisner (2008); diesel subsidies: GTZ.

Note: For the year 2004. △ = Oil supply >10 years of domestic demand and >30 million tons CO_2 emissions per year.

The benefits of electricity access far exceed the damage of associated emissions.

entirely on other poor people and were assessed at a carbon shadow price of $50 per ton of CO_2, and if no low-carbon alternatives were available, gross project benefits would be reduced by no more than $0.03 to $0.05 per kWh. And lower-carbon alternatives are available and the damages, if any, would be smaller.

Of course, people depend on energy in indirect ways, as for manufactured goods and employment. But figure 2.1 suggests that economic growth in the poorest countries generates little pressure on the atmosphere. The 50 least-developed countries, with a population of about 725 million, had energy-related emissions[5] of 121 million tons of carbon dioxide equivalent (CO_2e) in 2004, against the 12,949 million tons of CO_2e of the OECD. Since emissions are roughly proportional to income per capita, a 100 percent growth in the least-developed countries' income would generate about the same incremental emissions as a 1 percent growth in income in the OECD countries.

So there is no reason to impose any mitigation

burden on the world's poorest people. The energy access agenda could proceed independently of the mitigation agenda.

Nonetheless, there are important areas of connection between these two agendas. First, it is possible that carbon finance could support provision of electricity access through renewable energy. Second, price reform policies—which can have economywide emissions-reducing impacts—could help or hurt poor people, depending on how the reforms are implemented. This issue will be discussed at greater length in chapter 4. Third, policy choices, including pricing policies, can affect a country's long-term trajectory—that is, whether it follows the steep (emissions-intensive) or shallow path to wealth in figure 2.2. Finally, as the concept of energy access is broadened to include increased energy consumption by wealthier groups in middle- and upper-middle-income countries, growth begins to put more significant pressures on emissions. It is in these countries that the efficiency and pricing policies examined in chapters 5 and 6 offer the highest absolute levels of domestic savings and emissions reductions.

Chapter 3

Evaluation Highlights

- Since 1992, Bank operations have evolved an approach to climate change.
- Bank strategies have continually stressed energy efficiency and removal of price distortions.
- About two-thirds of Country Assistance Strategies in countries with high GHG emissions included a goal related to reductions, but only half of those that mentioned energy efficiency included specific objectives.
- Carbon accounting provides a way to balance the environmental costs and benefits of investments, but it should be approached with care.
- A systemwide approach is important to take account of trade-offs among economic and environmental costs and benefits.

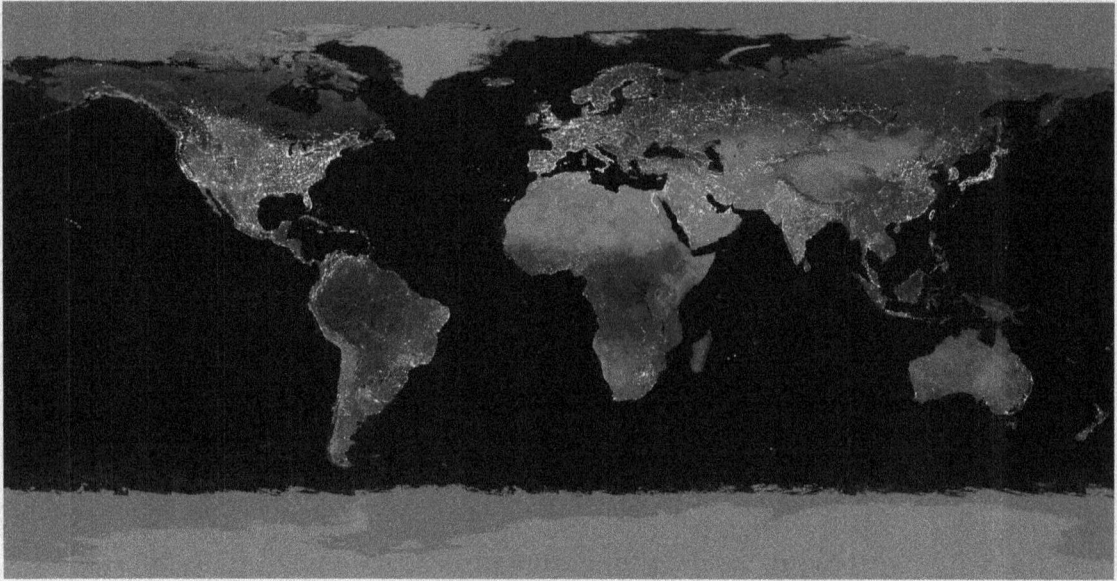

The lights of human habitation at night on Earth, as it would look from space with no clouds. This view of the Earth is a composite of many different satellite images. Photo ©NASA/Corbis, reproduced with their permission.

World Bank Operations and Climate Change

Until the 2008 announcement of its Strategic Framework on Development and Climate Change, the Bank Group had lacked a corporate approach to climate change. However, there has been scattered and increasing attention to GHG mitigation in energy and environment strategies and in country dialogue, and a growing portfolio of GHG-reduction and clean energy projects. This chapter briefly reviews relevant World Bank activities.

Climate in World Bank Policies and Strategies

Global attention to climate change surged during the 1980s and emerged with full force with the 1992 Rio Conference of the UN Conference on Environment and Development. The World Bank's *World Development Report* of that year highlighted the importance of addressing climate change (World Bank 1992). It pointed to ample scope for win-win policies, such as energy price reform and improvements in energy efficiency, but also noted the need to address environmental externalities through taxes or grants.

The win-win message was picked up in a 1993 policy paper, *Energy Efficiency and Conservation in the Developing World: The World Bank's Role* (World Bank 1993). The paper promised that the Bank would "continue its efforts toward increasing lending for components to improve EE [energy efficiency] and promote economically justified fuel switching." While only briefly mentioning GHGs, it outlined a four-point program to:

- Integrate energy efficiency issues into country policy dialogue.
- Decline to finance energy supply in the absence of structural reform.
- Give demand-side management (DSM) "high-level, in-country visibility."
- "Monitor, review, and disseminate the experience of new efficiency-enhancing supply-side and end-use . . . technologies . . . help finance their application; and encourage the reduction of barriers to their adoption."

The Bank's Energy Policy was published at about the same time and remains in force. It stresses "integrated energy strategies that help borrowing countries take advantage of all energy supply options, including cost-effective conservation-based supplies and renewable energy sources" as well as "cost-effective . . . options . . . to mitigate the negative environmental impacts of electricity supply and end use." It briefly mentions fuel switching and energy efficiency as means of abating CO_2 emissions.

Bank policies have included concerns about climate change since 1992.

About seven years later, four strategic documents reemphasized the themes of the 1992 *World Development Report,* while elevating the prominence of climate change. *Come Hell or High Water* (World Bank 1999) was concerned with climate change vulnerability and adaptation. It found that climate change risks were not well assessed in project preparation or in Country Assistance Strategies (CASs) and recommended attention to current and future climate variability.

In the late 1990s, strategic documents elevated the prominence of climate change.

Fuel for Thought: An Environmental Strategy for the Energy Sector (World Bank 2000) drew attention to mainstreaming energy into CASs and operations. It stated, "At the heart of mainstreaming environment within the Bank is the elimination of market distortions, particularly in energy pricing. As long as energy prices are subsidized or not at market level, and as long as gross interfuel pricing differences remain, it is difficult to formulate cost-effective measures to mitigate pollution from energy use."

With an emphasis on reducing the damages of local air pollution, *Fuel for Thought* stressed the need for a cross-sectoral perspective and proposed the use of "Energy-Environment Reviews" as an upstream analytic tool for promoting this perspective. One of the document's strategic objectives was to "mitigate the potential impact of energy use on global climate change." Its medium- and long-term outcome indicators for achieving this objective (through fiscal year 2008) include energy-efficiency programs in 10 states or countries; development of cleaner sources of energy (no quantitative goals); increasing the volume of energy trade among at least 6 countries; and doubling of power generation through renewable energy sources in at least 10 borrowers.

Reviewing post-1992 progress on this agenda, *Fuel for Thought* drew three main lessons:

That more time than initially estimated is needed to achieve results on environmental and social issues; that commitment is often missing on the part of the borrower to stay the course and to achieve real change; and that while there is strong engagement in the reform agenda, the strength of the Group's commitment to energy efficiency and the environment is not what it should or could be. The Group must substantially increase its efforts and improve its staff and skills mix if it is serious about implementing its principles in these areas.

The *World Bank Group's Energy Program* was presented to the Board of Directors and published in 2001. Although not a formal policy document, it reported that the "World Bank Group has set quantitative objectives for developing and transition economies to be reached by 2010." These included "reducing the average intensity of carbon dioxide emissions from energy production from 2.90 tons per ton of oil equivalent to 2.75" and "reducing the average energy consumption per unit of GDP from 0.27 ton of oil equivalent per thousand dollars of output to 0.24."

The *World Bank Group Environment Strategy* of 2001 dealt at length with the "threat posed by climate change to the development process." It continued to stress the twin themes of no-regret policies (including energy sector reform, energy efficiency, and fuel switching), together with continued collaboration with the GEF on renewables and use of the Prototype Carbon Fund (PCF) as a pilot to demonstrate the potential for carbon trading under the Kyoto Protocol. The strategy also pointed to mitigation opportunities in forestry and transport and promoted attention to mainstreaming efforts in climate adaptation. It stressed the use of Strategic Environmental Assessments, including Energy-Environment Reviews to ensure that local and global environmental issues are considered in the context of energy systems choices.

The World Bank Group reports that it committed $6.1 billion to renewable energy and $2.1 billion to energy efficiency during 1990–2004. In 2004, in Bonn, the World Bank Group made a commitment to expand its investments in new renewables (excluding large hydropower) and energy efficiency by 20 percent annually over 2005–09. Total reported commitments for new renewables were $860 million from fiscal 2005 to 2007, and commitments to energy efficiency were $952 million over the same period. According to data released by the Bank, the World Bank Group outperformed its Bonn commitment during 2005–07, committing about double its goal of $913 million.

In support of its commitments expressed through the Gleneagles Communiqué, "Climate Change, Clean Energy and Sustainable Development" (July 2005), the Bank developed an Investment Framework for Clean Energy and Development that was formally presented to the Development Committee in the spring of 2006; an Action Plan was endorsed by the Committee in spring of 2007. Inaccurately named, this evolving framework has three pillars: investment in power system expansion, with emphasis on increasing access for the poor; mitigation of GHGs from both energy and land use change; and adaptation to climate change. The mitigation component stressed energy efficiency as a "quick-win and high-payoff" pursuit, but focused on the mobilization of concessional funds for investments in clean technologies and the promotion of carbon trading.

Global Finance and Institutions

In the post-Rio era, the World Bank has been involved in the development of global institutions for climate change mitigation. These are briefly reviewed here for context.

The GEF was established in 1991 as the financial mechanism of the UNFCCC. The World Bank contributed to the design of the GEF's operational programs in climate change: removal of barriers to energy conservation and energy efficiency, removal of barriers and reduction of implementation costs for renewable energy, and reduction of the long-term costs of low-GHG-emitting energy technologies. The GEF has approved 634 climate change projects with grants totaling $2.3 billion and cofinancing of $14.6 billion. The GEF also supports interventions that increase resilience to the adverse impacts of climate change. It administers $300 million in three special funds, the Least-Developed Countries Fund, the Special Climate Change Fund, and the Strategic Priority for Adaptation Fund.

The 2001 Environment Strategy stressed no-regret policies and collaboration with GEF and the Prototype Carbon Fund.

The World Bank is an implementing agency of the GEF and, as such, has helped its client countries mobilize resources to cover the additional costs of initiatives aimed at meeting UNFCCC objectives. The World Bank Group GEF Climate Change Portfolio has evolved from mainly demonstration projects (that is, how to increase the efficiency of existing energy facilities and how to feasibly develop new and renewable energy sources) to a focus on market transformation in an effort to remove barriers to its present focus on mobilizing and enhancing the capacity of local financial markets to support environmental investments.

As it agreed to do in 2004, the Bank increased its commitments to renewables and energy efficiency.

Building on its experience with the set of pilot projects known as Activities Implemented Jointly, the Bank developed the Prototype Carbon Fund (PCF). The PCF was intended to pilot mechanisms for project-based GHG emissions reductions under the Kyoto Protocol. Already under development while the Kyoto Protocol was being negotiated, the PCF was formally launched in January 2000. The PCF was successful in raising funds and has since been supplemented by another 10 funds, all of them overseen by the Bank's Carbon Finance Unit.

The Carbon Finance Unit operates by identifying and financing emissions-reducing projects through agreements to purchase emissions reductions, for the most part destined for use under the Clean Development Mechanism (CDM).

By August 2007, the Carbon Finance Unit had raised a total of $2 billion. It signed purchase agreements of about $1.5 billion for 200 million tons of reductions from 89 projects (World Bank 2007a). This constituted about 20 percent of all transactions in the CDM. However, only 21 million tons had been issued through 2007.

About 11 percent of the portfolio is devoted to waste management (landfill-gas recovery), 7 percent to hydropower, 3 percent to biomass, 2 percent to wind, and 9 percent to energy efficiency. The portfolio is currently dominated (56 percent) by projects for the destruction of HFC-23, a potent GHG that is produced as a by-product of HCFC-22, a refrigerant that is a GHG and an ozone-depleting substance. HCFC-22 has been phased out in the developed world, but it is temporarily permitted for manufacture in the developing world.

Two-thirds of the 308 projects with climate change themes are in the energy/mining sector.

The impact and additionality of these projects, and the role of the PCF in shaping carbon market institutions, will be addressed in the subsequent phase of this evaluation. For current discussion, there are three noteworthy features of Carbon Finance Unit projects. The first is that, by design, they incorporate some form of carbon pricing. Second, they generally have no policy content. Carbon projects currently operate at a project or facility level, using payments for reductions as a way to make otherwise marginal projects bankable. Third, PCF/Carbon Finance Unit projects have mostly originated outside the Bank.

Mainstreaming

Projects

There are 308 World Bank projects with an explicit climate change theme; of these, 132 are IBRD/IDA investment loans, 10 are Development Policy Loans (DPLs), 86 are GEF, and 46 are carbon offsets (table 3.1). About two-thirds of these projects are mapped to the Energy/Mining Sector and include natural gas recovery, coal bed methane recovery, renewable fuel development, and energy conservation, among other activities. Projects mapped to the Environment Sector Board also involve energy efficiency and renewable energy. Rural development projects are mostly forestry related. These tallies include all projects with a climate change theme, regardless of the notional proportion dealing with climate. The thematic mapping, however, which is done by task team leaders, is not necessarily consistent or accurate.

Table 3.1: Climate-Themed Projects by Sector Board and Funding Source, Cumulative, 1990–2007

Sector board	DPL	IBRD/IDA	GEF[a]	Carbon offset	Other product lines[b]	Non-lending	Total financing
Education		3					3
Energy and mining	8	103	51	16	3	7	188
Environment	1	11	24	24	5	14	79
Economic policy					1		1
Financial sector	1	1		1			3
Private sector development			1				1
Rural development[c]		3	5	4	1	2	15
Transport		4	3				7
Urban development		4	2	1		1	8
Water supply and sanitation		3					3
Total	10	132	86	46	10	24	308

Source: World Bank data, June 2008.
a. GEF includes GEF and GEF medium-size projects.
b. Other product lines include guarantees, Montreal Protocol, special financing, and recipient-executed projects.
c. Investments in rural development and agriculture and rural development are combined under the section "rural development."

The number of projects has increased sharply since 2004, reflecting the entrance of the carbon funds (figure 3.1a). However, the total volume of climate-related components has stayed relatively constant since Rio (1992), except for a large spike in 2006 associated with the two large HFC-23 carbon projects (figure 3.1b). The other carbon projects are relatively small.

Country Assistance Strategies

A recent review by Nakhooda (2008) assessed 54 CASs issued over the period 2004–07 for mention of climate change. She found 32 that discussed GHG mitigation in a sectoral context, 18 with concrete targets related to mitigation. However, the quality and nature of these references varied. Some were concerned with CDM participation, while most did not mention GHGs explicitly. Only 11 mentioned climate vulnerability or adaptation finance.

For this evaluation, IEG reviewed the country strategies of the 33 Bank clients with the largest energy-related GHG emissions over the period 1995–2007 (table 3.2). Twenty of these countries

With the introduction of carbon funds, the number of projects increased sharply.

Figure 3.1: World Bank Climate-Themed Projects and Commitments (in $ millions by year, 1990–2007)

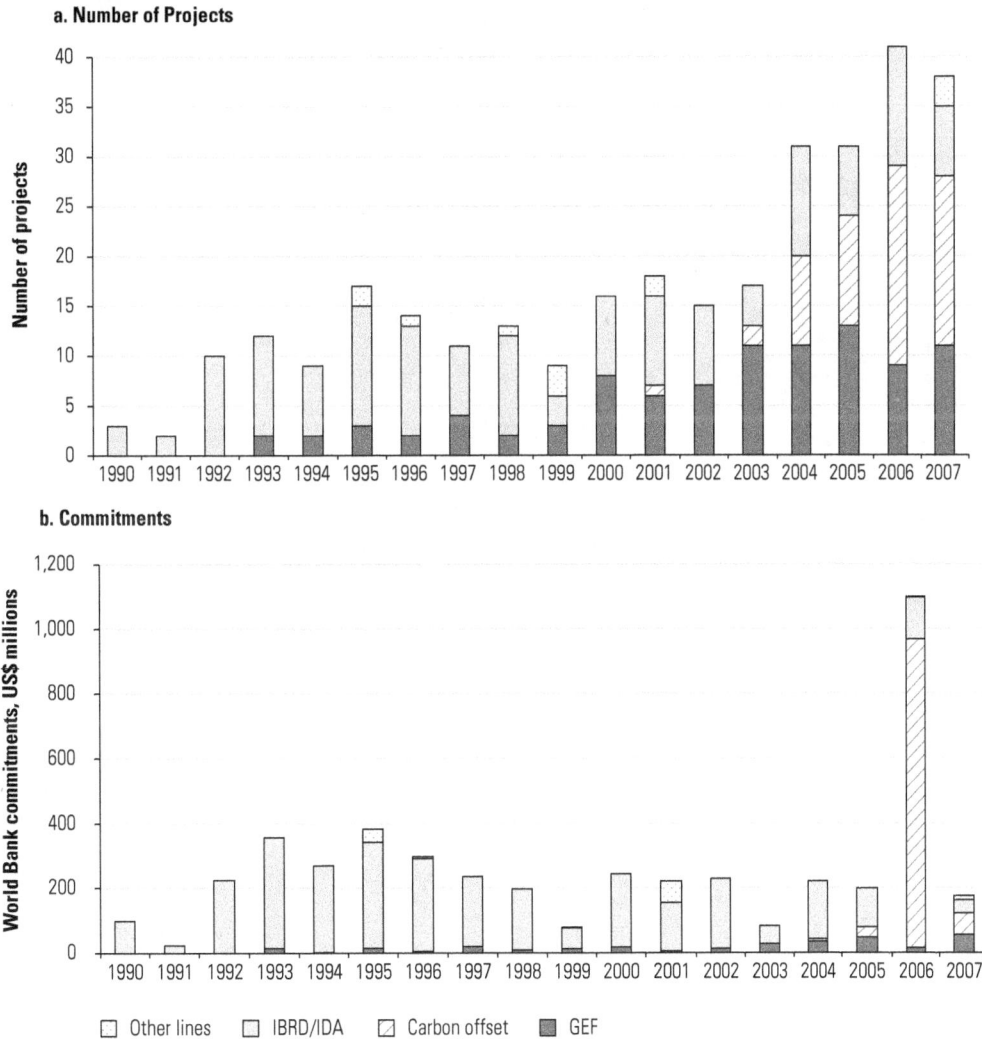

a. Number of Projects

b. Commitments

Other lines   IBRD/IDA   Carbon offset   GEF

Source: World Bank data, June 2008.

Note: Commitment amounts reflect proportion of total commitments associated with climate change; 2006 spike in commitments reflects two large carbon finance operations.

Table 3.2: CAS Goals for Energy Policies and Climate Change Issues, 1995–2007

Country	GDP per unit of energy use (2005)[a]	CO_2 emissions (2005)[b]	Power sector pricing policies	Primary energy pricing policies	Power sector reforms[c]	Primary energy sector reforms, including closing of loss-making coal mines	Efficiency policies and investments[d]
Algeria	6	88.10	0	1	1	1	0
Argentina	7	146.64	1	1	3	0	0
Azerbaijan	3	37.03	2	3	2	2	1
Bangladesh	7	39.82	4	3	6	2	0
Brazil	8	360.57	1	0	5	3	0
Bulgaria	4	50.54	2	3	5	3	1
Chile	7	66.19	0	0	0	0	0
China	3	5,322.69	2	1	7	1	5
Colombia	9	58.80	2	0	1	0	0
Egypt, Arab Rep. of	5	161.79	1	1	3	0	1
Hungary	8	59.84	0	0	1	0	0
India	5	1,165.72	3	1	4	1	1
Indonesia	4	359.47	4	3	3	4	0
Kazakhstan	3	198.01	2	0	2	1	0
Malaysia	5	155.51	0	0	0	0	0
Mexico	7	398.25	2	1	6	2	0
Nigeria	2	105.19	0	0	1	1	0
Pakistan	5	121.49	2	2	3	1	0
Philippines	6	78.06	4	1	5	0	0
Poland	6	284.64	0	0	3	5	2
Romania	5	99.34	3	2	5	4	0
Russian Federation	3	1,696.00	2	3	6	6	0
Serbia	na	52.56	1	0	0	0	1
Slovak Rep.	5	37.81	1	1	1	1	0
South Africa	3	423.81	0	0	0	0	0
Thailand	4	234.16	0	0	2	0	1
Turkey	7	230.04	0	0	4	4	0
Turkmenistan	na	49.64	0	0	0	1	0
Ukraine	2	342.57	3	1	4	4	1
Uzbekistan	1	117.97	2	4	2	2	0
Venezuela, R.B. de	4	151.29	0	0	1	1	0
Vietnam	4	80.38	5	0	4	1	2

Note: na = not available.

a. GDP per unit of energy use, 2005 purchasing power parity $ per kilogram of oil equivalent. *World Development Indicators 2008*, 2005 data table 3.8, p.168.

b. Energy Information Administration. World Carbon Dioxide Emissions from the Consumption and Flaring of Fossil Fuels, 1980–2005 (Million Metric Tons of Carbon Dioxide). International Energy Annual 2005 Table Posted: September 18, 2007. http://www.eia.doe.gov/iea/carbon.html.

c. Including regulatory agency setup and reform.

d. Includes research, development, demonstration, and planning for energy efficiency, standards and certification, mandates and incentives for DSM promotion, marketing awareness of energy-efficient technologies to support DSM, investments in DSM and supply-side efficiency.

e. Includes incentives for use of renewable energy or clean fuels, markets for grid-connected renewable energy; sharing of power from renewable energy; investments in hydropower, wind, biomass, and other types of renewable energy.

f. Energy efficiency identified in CAS document as an overall objective for the energy sector.

Renewable policies and investments[e]	Energy-efficiency goal (high-level target)[f]	Energy security (fuel mix; alternative sources of energy)[g]	CAS goals on GHG reduction, global treaties on climate change and ozone-depleting substances[h]	Share of energy-efficiency component in total spending for energy, %[i]	Share of energy-efficiency projects in energy portfolio, %[j]	Total energy commitments,[k] US$mln
0	0	0	1	0.0	0.0	126.34
1	0	0	6	0.0	0.0	557.81
0	0	0	0	0.0	0.0	125.94
2	1	1	0	0.0	0.0	919.84
0	4	0	4	1.0	3.6	1,558.93
1	4	0	2	25.3	27.3	191.60
2	1	1	1	30.6	14.3	22.84
1	9	5	5	5.7	15.9	8,138.29
0	1	0	1	0.0	0.0	321.76
0	0	0	1	0.0	0.0	650.55
0	0	0	0	0.2	33.3	235.70
0	2	1	1	0.5	6.8	6,243.18
1	2	1	1	0.4	4.8	2,525.44
0	0	0	0	0.0	0.0	427.49
0	1	1	0	0.0	0.0	200.00
0	3	3	6	1.5	8.0	1,027.41
0	0	0	0	1.4	14.3	624.45
0	1	1	2	0.0	0.0	2,327.04
2	2	1	1	0.1	4.8	1,410.35
0	2	0	0	31.9	41.2	1,295.97
1	2	1	1	2.3	12.5	879.72
0	3	1	5	13.7	31.6	3,291.09
0	0	0	0	45.0	25.0	107.06
1	1	0	1	0.0	0.0	0.00
0	1	0	2	0.0	0.0	3.34
0	1	0	1	0.9	13.3	1,337.55
1	0	2	0	0.0	0.0	2,391.62
0	0	0	0	0.0	0.0	2.25
0	4	2	3	18.8	17.6	1,310.39
0	0	0	0	0.0	0.0	0.63
0	0	0	0	0.0	0.0	9.00
5	3	0	1	0.7	11.1	1,460.71

g. Energy security identified as a high-level goal, includes also the objectives for fuel mix improvement and search for alternative energy sources.

h. CAS objectives related to CO_2 reduction, ratification/discussion of Kyoto and Montreal Protocols, interest, and priority of climate change issue.

i. Energy efficiency components' share based on World Bank 2005c, 2005e, 2006b; World Bank and IFC 2007, and total energy spending in 1990–2007. Excludes IFC and MIGA.

j. Share of projects with energy efficiency components based on World Bank 2005c, 2005e, 2006b; World Bank and IFC 2007, and total number of Bank Group renewable energy and energy efficiency reports. Excludes IFC and MIGA.

k. World Bank data. Energy commitments represent the commitments only for energy sectors.

had at least one strategy with overall goals related to GHG reduction, the UNFCCC, or the Montreal Protocol. But emphasis was uneven among countries. The greatest attention was given to these goals in Argentina, China, Mexico, and Russia.

Of 33 CASs for the Bank clients with the largest energy-related GHG emissions, 20 had a strategy with a goal related to GHG reduction.

Table 3.2 also tabulates CAS goals related to some of the potential win-win policies discussed later in this report. The tabulation includes statements of high-level goals and specific, potentially monitorable objectives. For 20 of the countries there was some mention of energy efficiency as a high-level goal. However, in only 10 countries were specific goals mentioned. This suggests a disconnect between rhetoric and action. But of the 10 countries singled out, 7 were in the most energy-intensive half of the group.

For the 20 countries with CASs that mentioned energy efficiency, only 10 included a specific goal.

In this set of countries, 17 had specific goals related to primary fuel pricing, 21 had power-pricing goals, and 25 had goals related to power sector reform. The outcomes of some of these goals will be examined in chapter 4. In sum, taking the Bank's country strategies as strong indicators of country interest in these agendas, such interest is widespread but not universal among client countries.

Cross-Sectoral Analyses

IEG's *Environmental Sustainability: An Assessment of World Bank Group Support* (IEG 2008b) stressed the need for cross-sectoral integration of environment and infrastructure concerns. As noted above, cross-sectoral analysis was emphasized in 2000 by *Fuel for Thought* and in the 2001 *Environment Strategy,* which pointed to Strategic Environmental Analyses (SEAs) and Energy-Environment Reviews (EERs) as instruments for accomplishing this. SEAs are the sectoral or policy generalization of project-level environmental impact analyses, which comprehensively assess the costs and benefits of alternative plans, taking environmental externalities into account. A related tool is the Country Environmental Analysis (CEA), intended to mainstream environmental issues into overall country strategic planning. SEAs

and EERs were introduced around 2000, while CEAs started in 2003.

Full-scale EERs have been completed and published for Bulgaria, Egypt (incorporated in the subsequent CEA), Iran, Mexico, and Turkey. The Egypt, Iran, and Mexico EERs address the countries' large fuel subsidies, finding them a major source of health damage as well as fiscal drains.

Iran's fuel subsidies in 2001 were estimated at 17.8 percent of GDP, and the damage of air pollution to health was estimated at 8.4 percent of GDP and growing. The report explored scenarios for price reform and additional sectoral measures, finding that price reform would cut health damage in half, though at some cost in inflation.

The Egypt CEA similarly found that adjusting energy prices to opportunity cost levels would reduce local damage by $200 million yearly, with additional savings available from implementing win-win efficiency measures.

The Mexico EER found that removal of power subsidies would reduce CO_2 emissions by about 3 percent. There would be a very small (<0.1 percent) negative effect on the welfare of the poor, which could easily be compensated through subsidy savings.

It is difficult to trace impacts of these studies. Egypt and Iran have increased fuel prices, though their prices remain well below world levels. Mexico has increased its level of fuel subsidies, but partnered with the Bank on a climate-oriented DPL.

SEAs have been used to systematically assess hydropower options. Hydropower has been contentious because of its potential for environmental damage and social disruption. While generally considered to have low GHG emissions, hydro plants with anoxic tropical reservoirs can emit methane. In principle, a comprehensive multi-attribute assessment of all options is superior to an environmental impact assessment of a predetermined and possibly suboptimal site. An SEA for Nepal (Government of Nepal 1997) ranked 138 potential medium-size hydropower projects for economic, environmental, and social impact; it prioritized 7 as having low impact. One of the seven was chosen for finance under the subsequent Nepal Power Development Project. An SEA for the Laos hydropower sector (Norplan 2004) assessed 21 proposed hydropower sites. While it notes that some threaten primary forests, and calculates environmental costs ranging from $0.001 to $0.136 per kWh, it does not propose a ranking or do a trade-off analysis. Finally, an SEA for the Nile Equatorial Lakes Region (SNC Lavalin International 2007) screens hydropower options (and thermal alternatives) against economic, environmental, social, and risk criteria, including life-cycle CO_2 emissions, although it does not take account of methane emissions. It discusses a variety of scenarios for sector expansion and considers a complex set of trade-offs.

According to the recently released World Bank assessment of CEAs (Pillai 2008), as of early 2008 the Bank had initiated 25 of the analyses. Of the 16 completed CEAs, those on Belarus, Egypt, and India mentioned climate change in the context of energy policies. These sectors were treated also in Bangladesh, Colombia, Pakistan, and Serbia-Montenegro. The India CEA dealt at length with the relationship between coal power and air pollution. It emphasizes the promotion of energy efficiency and renewable energy, including finalization of the Renewable Energy Policy and support for upgrading inefficient old coal plants. (A contemporaneously prepared IBRD/GEF project proposes to provide such support.) However, the CEA is silent on the well-known energy-irrigation nexus: the poorly targeted electricity subsidies that encourage unsustainable use of scarce groundwater.

The Belarus (2002) and Serbia-Montenegro (2003) CEAs emphasize energy efficiency and the need to rationalize prices and reduce subsidies. Tariffs did rise in Belarus over 2002–05. An efficiency project is in the pipeline for Belarus, and several projects with efficiency components

have been initiated in Serbia-Montenegro. The Bangladesh, Colombia, and Pakistan CEAs discuss vehicle-related emissions and fuel quality, but do not link the discussion to broader energy or transport issues. The Bangladesh CEA, for instance, discusses pollution from diesel engines but does not discuss the role of diesel subsidies. This is in contrast to the Egypt CEA.

SEAs have been used to systematically assess hydropower operations.

Strategic Considerations for the Bank: Accounting for Local and Global Impacts

Policies and projects supported by the World Bank Group have both local and global effects. Not all of them are win-win. In the Bank Group's country-based model, infrastructure investments, including those in transport and power, are seen by many clients as an important source of growth and poverty reduction. Support for growth, rather than climate change mitigation, remains the focus of the Bank Group's energy support, even as it moves to increase the share of renewable power within that support. As it allocates its efforts and funds across activities, should the Bank Group take into account their global impact on climate as well as their local impact on welfare? Without presuming to answer that question, this evaluation looks at methods for assessing the trade-offs and complementarities, rationales for using them, and experience in their application.

Of 16 completed CEAs, 3 mentioned climate change in the context of energy policies.

There are divergent opinions on whether and how carbon emissions should enter into project analysis and selection. One view holds that since developing countries have not taken on responsibility for emissions reductions, emissions should not be a consideration in project selection, except where emissions reductions are a source of revenue. An opposing view holds the Bank responsible for emissions it finances, in the same way that private companies are beginning to view carbon emissions as liabilities.

A third view sees valid differences in scope between the concerns of the Bank Group and those of any individual developing-country client.

A Bank-supported project in one country could damage or benefit other client countries.

This reflects the underlying tension between the Bank Group's country-based model and its support for global public goods, an issue discussed at length in IEG's 2008 *Annual Review of Development Effectiveness.* The client is properly interested in promoting its own development, does not accept limits on emissions, and correctly considers that its historical or per capita contribution to global emissions is small relative to that made by developed countries. The Bank, however, is concerned with the welfare of all its clients and with climate risks to its global portfolio. From the viewpoint of environmental economics, a Bank Group–supported project in one country may, at the margin, accelerate or retard climate change, and thereby damage or benefit other vulnerable client countries. These marginal impacts are in addition to the much more substantial damages from the cumulative emissions of developed countries.

Carbon Accounting at the Investment Level

How might the Bank take emissions into account in project design and selection? Some observers advocate proscribing all funding for coal power plants, oil extraction, or other fossil fuel-related activities. A provocative analogy would be the Bank's policy toward tobacco. Tobacco is a remunerative export crop that provides domestic poverty-reduction benefits through employment. But it is also an addictive substance that imposes substantial transborder economic and health costs. So Bank investments in tobacco pose a trade-off between local benefits and global damages. Recognizing this, in 1991 the World Bank adopted a policy prohibiting lending or investments in tobacco production, processing, or marketing. (However, the policy allows for exceptions in countries where tobacco represents more than 10 percent of exports.)

Carbon accounting provides a nuanced way to balance the climate costs and benefits of investments.

Carbon accounting—estimating and monitoring project emissions—provides a more nuanced way to balance the costs and benefits of investments. It can be used to assess alternative ways of fulfilling a particular project objective, such as constructing a 200-megawatt power plant. Alternative technologies, such as coal and geothermal, could be compared on purely economic criteria as well as on carbon emissions. Four uses have been suggested for this information.

- First, carbon accounting may promote more analytic rigor and uncover win-win project alternatives with higher returns and lower emissions.
- Second, carbon accounting may be used to justify carbon market finance. If the coal plant is cheaper than the geothermal plant but emits more CO_2, it is possible to compute the value per ton of CO_2 reductions at which the geothermal plant becomes more attractive (the switching cost). If emissions reductions can be sold at this price on the carbon market, then the cleaner plant can be funded. Note, however, that the long-term and large-scale availability of carbon finance is uncertain, pending the outcome of negotiations on the global climate regime.
- Third, carbon accounting provides information on the switching cost. This information is useful in assessing the impact of future policies on emissions and on the economy. It can inform models used by climate scientists, negotiators, and others.
- Finally, and most controversially, a shadow price—representing the marginal impact of a change in emissions—could be applied to a project's emissions, and this impact incorporated in an economic rate of return or cost-benefit analysis. These, in turn, could enter project appraisals or evaluations, as is often done with other kinds of environmental externalities.

Carbon shadow pricing is not new to the Bank. In 1999, the Bank published a pilot study (ESMAP 1999) that examined how carbon shadow pricing might affect project choice. It found that 41 percent of loans examined would become uneconomic if their gross emissions carried a shadow price of $11 per ton of CO_2; this proportion was lower if emissions were netted against a business-as-usual baseline. But among the eight thermal plants examined, negative switching values (that is, apparently overlooked win-win alternatives) were found for six. The study found no barriers to calculating carbon footprints in the projects it assessed, which were well-defined generation projects. Even in the absence of carbon markets, it recommended shadow pricing as an informational practice that might uncover cost-effective switches.

Carbon shadow-pricing is often incorporated in GEF and other projects that have emissions reduction as a cobenefit. For instance, the efficiency projects and subprojects described in box 5.1 had returns of up to 289 percent when carbon benefits were included. Carbon pricing and monitoring is already a feature of project design and appraisal for the Bank's carbon projects. Under the CDM, carbon projects must estimate, and then verify, actual emissions. These emissions are compared to business-as-usual emissions to assess project impact—that is, to quantify emissions reductions. Financial appraisal takes into account the value of emissions reductions—an actual, rather than a shadow, price of carbon. A standard procedure for justifying a carbon project is to argue that the carbon price is greater than the switching price.

An important feature of CDM projects, including those of the World Bank, is that they require rigorous independent monitoring and verification of emissions. This information is published through the CDM and provides a public good: rapid feedback on the outcomes of new types of clean technology. For instance, through this reporting process it has rapidly become clear that projects involving landfill-gas recovery (generation of power from municipal waste) are consistently underperforming compared with appraisal projections. This is prompting re-examination of the engineering models used to predict project output.

IFC's recently adopted Performance Standard 3 requires its corporate clients to annually quantify direct and some indirect[1] GHG emissions for projects that are expected to emit more than 100,000 tons of CO_2e annually. Clients are also required to evaluate "technically and financially feasible and cost-effective options to reduce or offset project-related GHG emissions," including carbon finance, changes in project design, and emissions offsets.

This monitoring and assessment requirement—which has no counterpart in the World Bank's safeguard policy—is a step forward in disclosure and transparency and will provide lessons for the World Bank and other funders. It will stimulate scrutiny and discussion of project alternatives. As an early example of the standard's application, consider the IFC's environmental assessment for the Lanco Amarkantak thermal power plant, which is scheduled to emit 4.2 million tons of CO_2 per year. In addressing Performance Standard 3, the publicly disclosed environmental review represents the plant's emissions intensity of 910 gCO_2/kWh as better than the Indian national average of coal plants at 1,225 gCO_2/kWh.[2] However, the relevant comparison is to new coal plants, rather than the existing semi-obsolescent fleet as a whole. Modern subcritical coal plants emit 855–880 gCO_2/kWh (see box 2.1), though Indian levels may be higher because of differences in coal quality. Alternatively, one could compare the plant's performance to the systemwide build margin across all energy types, which is 680 gCO_2/kWh according to the Central Electricity Authority.[3] However, the plant's environmental statement says that the owner will explore various means of reducing emissions, including afforestation offsets and cofiring with biomass.

Carbon shadow pricing dates back to 1999 in the Bank.

IFC's recently adopted Performance Standard for GHG emissions is an important step, for which there is no counterpart in the Bank.

Monitoring and reporting emissions could stimulate discussion of project alternatives.

Carbon Accounting at the System Level

As the review of SEAs and EERs pointed out, any individual power plant is a small component in an interconnected energy system. Many of the important trade-offs among economic and environmental costs and benefits occur at this system level. So an analysis of choice of technology for a predetermined goal at a predetermined location may completely miss the crucial systemwide options. For instance, systemwide power efficiency improvements might substitute for a new generating plant. At the same time, it is possible that a proposed fossil fuel plant, emissions-intensive when considered on its own, is a necessary part of a portfolio that includes lower-carbon sources. This suggests a system-wide approach to assessing costs and benefits, including emissions.

Again, this approach is not new to the Bank. Two state-level studies of the Indian power sector, while not formally designated as EERs, exemplify the approach and are noteworthy for monetizing local and global environmental damages and assessing trade-offs against the consumption value of electricity. The studies of Rajasthan and Karnataka (ESMAP and others 2004a, 2004b), undertaken in collaboration with the state governments, examine supply and demand alternatives—including power sector and tariff reform—for meeting the states' power needs. Environmental impacts include three kinds of local air pollution (SO_x, NO_x, and PM_{10} [particulate matter of 10 micrometers or less]), consumptive water use, and CO_2 emissions, with damages put at $55 per ton of CO_2e.

A systemwide approach has been studied in

The Rajasthan study shows that failure of reform, by choking off power capacity expansion, severely stunts economic performance and leaves local pollution virtually unchanged as industry switches to small, polluting diesel generators. Stunted growth leads to slightly lower CO_2 emissions, but the extreme implicit cost of this reduction—$480 per ton of CO_2e—easily rules out policy failure as a climate mitigation strategy. At the same time, going from a basic reform scenario to one that includes some degree of tariff rationalization and DSM is win-win. It boosts the value of the investment program by 13 percent and reduces air pollution and water use by about 6 percent, and CO_2 emissions by about 4 percent.

The South East Europe Generation Investment Plan computes a least-cost power expansion plan for the region (2005–20) under a number of scenarios that comply with European Union (EU) environmental standards for air pollutants, including CO_2. Table 3.3 shows that imposition of a €10 per ton

And it has been applied in South East Europe.

Table 3.3: Effect of Carbon Shadow Price on Generating Capacity Mix for South East Europe, 2020

CO_2 shadow price	Lignite plus coal (%)	Gas (%)	Nuclear (%)
€0	36.9	13.1	10.3
€5	34.0	16.0	10.4
€10	30.9	17.2	12.3

Source: South East Europe Consultants (2005).
Note: The baseline is a cost-minimizing optimal scenario, which features less rehabilitation of old plants than the official scenario. Other elements of the generating mix, including hydropower, are constant across scenarios.

of CO_2 shadow price shifts 6 percent of total capacity from lignite and coal to gas and nuclear. Even at €10 per ton, the optimal plan still involves the construction of large new lignite-fired plants in Kosovo. However, since the plan was generated, the European Trading System for carbon has started operation, and CO_2 traded at €25–30 in mid-2008.

The Role of Economic Analysis of Projects

Among practitioners of carbon shadow pricing, there is a debate on what price level to assign. This value, which represents the damages imposed by an additional ton of CO_2, is set in the *Stern Review* at $85 per ton of CO_2e; the U.K.'s Department for Environment, Food, and Rural Affairs recommends a value of £26 per ton of CO_2e for project appraisal (rising over time).[4]

However, while this debate was going on, the price of oil, gas, and coal rose drastically (see figure 3.2). The mid-2008 price of oil was equivalent to the 2006 price of oil plus a $135 per ton CO_2 price. Although prices have since declined, expected future prices remain high by recent standards. Hence, actual project appraisal decisions should already be moving in directions similar to those suggested by carbon shadow pricing.

For project appraisal to send these signals, it must value energy and electricity at economic prices. This is not easy in systems where prices are distorted or where electricity supply is constrained. For instance, the Rwanda Emergency

Electricity Project values additional electric output at $0.15 per kWh, even though this is below the cost of provision, and far below the likely willingness to pay. And correctly valuing additional electricity access is a technical problem that requires information or assumptions about demand, and the IEG review of rural electrification (IEG 2008d) found that only 5 of 13 projects examined used best-practice techniques. Similarly, distortions in coal and (especially) gas markets need to be accounted for in project appraisal.

Moreover, economic analysis should incorporate allowance for energy price volatility. Because fossil fuel prices are volatile and uncorrelated with variation in wind and rain, investments in renewables and energy efficiency carry a risk-hedging benefit—in effect, another kind of shadow value. Some ESMAP work has been important in drawing attention to this (Hertzmark 2007). But while the carbon shadow value is perceived only when carbon markets are active, the risk-hedging value associated with renewables is a clear benefit to the investor or to the host nation—depending who bears the price risk.

Furthermore, most development and carbon impact assessments look within the boundaries of the project. But projects—especially low-carbon projects—often aspire to catalyze replication and diffusion through demonstration or market transformation effects. These include learning-curve effects, reduction of perceived risk, and stimulus of supply and service markets. Multipliers should therefore attach to both the development and carbon effects of these projects. In practice, spillover effects may dominate within-

High energy prices act like carbon taxes in some ways.

For project appraisals to send the right signals they must value energy and electricity at economic prices, but this is not easily done.

Renewable energy and energy efficiency are hedges against volatility of fossil fuel prices.

Figure 3.2: Real Energy Prices of Coal, Gas, and Oil, 1990–2008

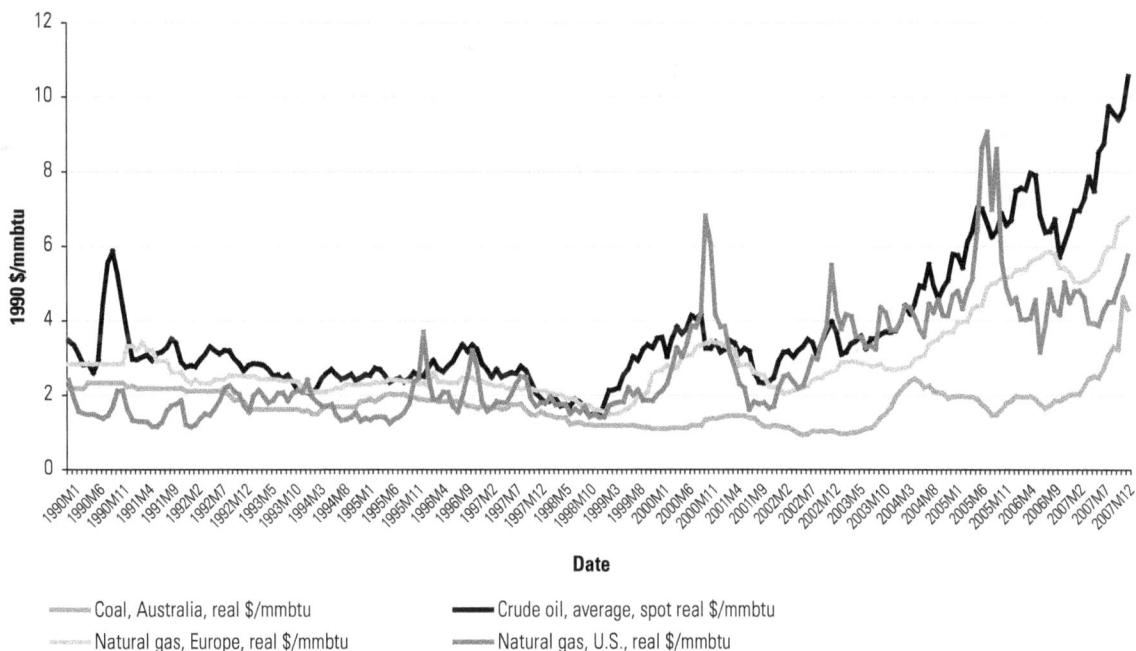

Date

Coal, Australia, real $/mmbtu
Crude oil, average, spot real $/mmbtu
Natural gas, Europe, real $/mmbtu
Natural gas, U.S., real $/mmbtu

Source: World Bank Global Economic Monitor.

Appraisal should consider spillover and demonstration effects.

project impacts. Consideration of spill-over effects in project selection and appraisal would tend to bring efficiency and renewables projects to much greater prominence.

Carbon Accounting at the Level of the Bank Group

While measuring a project's gross emission may be straightforward, the net impact on emissions could be very different.

The Bank needs to recognize that pursuing its primary mission of poverty reduction will inevitably put upward pressure on global emissions, simply because people with rising incomes demand more energy, and more agricultural products that will compete for land with forests. While provision of basic energy access to the poorest will have little aggregate impact on emissions, policies stimulating robust, shared growth in the developing world will indirectly spur emissions in rough proportion to income. These pressures can be moderated or exacerbated by Bank-assisted policies that shape energy and land use.

Carbon accounting (or footprinting) of Bank operations should be approached with caution. An advantage of footprinting is its ability to focus attention and stimulate critical and creative thinking on emissions reductions. (See, for instance, the success of carma.org, which reports on worldwide emissions of all power-generating plants.) However, to the extent that footprinting is not comprehensive in scope, it could be misleading or even lead to perverse outcomes. And it is important to note that the Bank Group's current ability to quantify aggregate impacts on other aspects of development is limited.

Bank-supported policy reforms could have larger impacts than investments.

One problem—which also applies to project-level carbon accounting—concerns measuring a project's GHG impact. It is relatively straightforward to measure gross emissions. But the net impact on emissions could be very different. Gas and combined heat and power plants are large gross emitters of CO_2. But where these plants substitute for coal-fired power, they could realize very large reductions in emissions. Emphasis on gross footprints might discourage such win-win investments. At the same time, net footprints have to be reckoned against a counterfactual: what would have happened in the absence of the Bank project? These counterfactuals are potentially subject to manipulation. This possibility has become a lightning rod for criticism of the CDM (which also uses such counterfactuals) and will be considered at greater length in the second phase of this evaluation.

A second and even more fundamental problem is that Bank-supported policy reforms could easily have impacts (positive or negative) that swamp investment-level footprints. Thus footprinting efforts, if undertaken, should be carefully qualified as to scope and methods.

Taken together, these considerations suggest a multilevel menu of options related to carbon accounting. First and most basic projects should employ rigorous economic analysis in appraisal, using economic values for fuel and power prices and taking price volatility, local environmental externalities, and demonstration effects into account. Second, the Bank could undertake carbon accounting at the project level, computing switching values for high- and low-carbon alternatives. Publication of these analyses would inform the global community about the costs of carbon abatement and would be an important public good. Third, the Bank could support interested clients in creating energy system expansion plans that take environmental impacts into consideration. These could be used to validate that proposed investments were consistent with economic and national environmental priorities. With the Bank's supporting role defined, these plans could also be used to provide a more comprehensive measure of the Bank's impact on emissions.

Box 3.1: The $135 per Ton CO_2 Price Is Already Here

Just a few years ago, climate policy scenarios controversially envisioned a world of universal high carbon taxes in the 2030s. The mid-2008 world bore a striking resemblance to those scenarios.

Plans for climate change mitigation usually include some provision to attach a real or implicit price to GHG emissions. The proposal is a mainstay of environmental economics: GHGs impose widespread costs on the environment, so those costs should be internalized in people's decisions on burning fuel, clearing forests, and so forth. This would balance costs and benefits in the short run, and motivate research and development toward cleaner technologies over the longer run. CO_2 prices could take the form of taxes on emissions, a requirement to buy an emissions permit, an opportunity to sell emissions reductions, or a combination of these measures. Thus, pricing CO_2 does not necessarily entail a tax on developing countries.

Much debate and analysis have been devoted to assessing carbon prices that would advance GHG stabilization goals, and yet be politically feasible. Various global models of mitigation, for instance, require CO_2 prices of $30 to $275 in 2020 (rising over time) to stabilize atmospheric concentrations at 550 ppm (Clarke and others 2007). Questions about the acceptability of this kind of price level underlie much of the negotiation on the global climate regime. Carbon financiers have explored the impact of certified emissions reductions (carbon credits) at $5 to $10/ton of CO_2 on investments in clean energy. And there is an ongoing debate about whether to incorporate carbon pricing in the World Bank's investment analysis. (That is, in assessing a project's benefit-cost ratio, should global damages attributable to GHG emissions be included in the cost?) Meanwhile, skyrocketing energy prices provide a taste of what carbon prices would feel like for consumers of fossil fuel. The table below shows the equivalence between a carbon price and a fuel price increase for three fuels. Suppose fossil fuel prices had remained at their 2006 levels. The table shows the CO_2 price (or tax) equivalent that would equate consumer prices to observed May

2008 levels. For instance, the May 2008 petroleum price is equivalent to that of 2006, with a $135 per ton of CO_2 tax added. The CO_2 price equivalent differs among fuels because of their different carbon content.

The table provides food for thought. First, mid-2008 price levels give some indication of the impact of high carbon prices in a scenario where global energy prices subside. Second, reactions to those prices give some indication of the short-run scope for adjustment to carbon prices and the implications for carbon emissions. In the United States, for instance, there has already been a sharp drop in sales of fuel-inefficient vehicles and an increase in ridership in public transportation. Third, the equivalence provides a new perspective on carbon shadow pricing of investments. There have been objections by developing countries to the use of carbon shadow pricing in the investment decisions of multilateral development banks as an unwarranted imposition of responsibility for emissions. However, prudent investment decisions should account for the possibility that energy prices will stay high (or spike high) during the life of a project. This self-interested calculation, focusing entirely on energy price volatility and not on carbon, will tend to favor renewable energy and energy efficiency in much the same manner as would a carbon shadow price. It does not obviate the burden of financing these investments.

However, note that the analogy between energy price hikes and carbon prices is imperfect in several important respects. First, the differing carbon and energy contents of fuels mean that a true carbon price would fall most heavily on coal, inducing substitution of other fuels. Second, energy price hikes, unlike a carbon price, would encourage the development of nonconventional sources of fossil fuels, including highly emissions-intensive sources such as oil shale and tar sands. Third, a global carbon price would depress the supply price of energy, so that its effect on final prices would be somewhat muted. Finally, the distributional consequences of carbon taxes, carbon permits, and high energy prices are quite different.

	Australian coal $/ton	Crude oil, avg. spot $/bbl	LNG (Japan) $/mmbtu
Mean price, Jan-Dec 2006	$49.09	$64.29	$7.08
Price, May 2008	$131.00	$122.63	$11.90
$/ton$CO_2$ equivalent of the 2006–08 fuel price increase	$31.77	$135.08	$80.52
Notes:			
Equivalence of a $1/ton CO_2 price on commodity price in physical units	$2.58	$0.43	$0.06
Equivalence of a $1/ton CO_2 price on commodity price per energy unit (mmbtu)	$0.103	$0.074	$0.060

Source: IEG calculation based on Development Prospect Group "Pink Sheet" (at http://www.worldbank.org/) commodity price data.

Note: Bbl = barrel, mmbtu = millions of British thermal units.

Chapter 4

Evaluation Highlights

- Subsidies are a large but poorly monitored drag on developing-country economies—removing them would increase economic efficiency and reduce GHG emissions.
- In countries where taxation has kept fuel prices high, emissions are lower.
- Most subsidies go to better-off consumers.
- Subsidy reduction can fund social protection that is better targeted to poor people.
- Power price reform goals have often been achieved, especially in transition countries.

Chapter 4

A resident of Palu, Indonesia, receives money under a cash transfer program instituted by the government to cushion the impact on poor people of a reduction in fuel subsidies. Photo by Basri Marzuki, reproduced with his permission.

Subsidies and Energy Pricing

Energy subsidies hobble economies, spur GHG emissions, and benefit primarily the better-off. While the record energy prices of 2008 underline the lose-lose nature of most subsidies, the drawbacks of energy subsidies are a longstanding concern. The solution seems obvious: rationalize prices and use the savings to provide more effective social protection for the poor and vulnerable. But like most apparently win-win propositions, it is not easy to put into practice. This chapter reviews efforts to do so.

The Nature of Subsidies and Price Distortions

Subsidies and price distortions take many forms and can be difficult to measure with precision (Morgan 2007; UNEP 2003). The most obvious are on-budget payments by governments to producers or consumers of energy. However, many subsidies are off-budget, and therefore harder to detect and calculate. Oil- and gas-producing countries often sell these fuels to consumers at a price below their economic value. The forgone revenue or opportunity cost constitutes a subsidy to buyers. Similarly, electricity is sometimes sold to consumers below the short-run marginal cost, and often below the long-run marginal cost. Assessing these implicit subsidies requires accurate estimation of the economic values involved. There can also be direct capital subsidies or tax benefits for energy producers.

The Problem with Subsidies

First, energy subsidies are enormous. Despite considerable progress in policy reform, there are still large subsidies to gas and oil outside the OECD, and substantial remaining coal subsidies within the OECD. These subsidies are not regularly, comprehensively, or consistently monitored. But ad hoc surveys show them to be huge. The International Energy Agency estimated that there was about a quarter-trillion dollars of annual consumption subsidies for electricity and fossil fuels outside the OECD in 2005 (IEA 2007). The largest subsidizers in absolute terms were Russia, Iran, Saudi Arabia, India, Indonesia, Ukraine, and Egypt—all with more than $10 billion a year in subsidies.[1] Implicit subsidies for gas and oil play a large role.

The OECD has about €29 billion in subsidies, mostly to energy producers (European Environment Agency 2004, quoted in Morgan 2007). Developed-country subsidies for biofuels are increasingly important. However, subsidies are poorly monitored and take a variety of forms. For instance, public spending preferences for roads versus urban transit or long-distance rail

Subsidies, though large, are not regularly, comprehensively, or consistently monitored.

provides an implicit subsidy for more emissions-intensive transport. Thus, the total scale of energy subsidies may be extremely large.

Not included in these estimates is theft of or nonpayment for electricity. These effectively act as subsidies to nonpaying users, many of whom are poor, but some of which are large farms, enterprises, or government entities (Smith 2004). A rough guide to the magnitude of these subsidies is provided by statistics on transmission and distribution losses, both technical (physical) and nontechnical. Purely technical losses are likely to be less than 15 percent, so excessive losses suggest theft. Reported transmission and distribution losses exceed these rates in many countries, including Ecuador (43 percent), Moldova (38 percent), India (31 percent), and Pakistan (24 percent).[2]

They are a huge drag on the economy and the public purse in some countries.

Second, *in some countries subsidies are a huge drag on the economy and on the public purse.* In Egypt in 2006, for instance, energy subsidies were about 12 percent of GDP—a bit more than half on budget, the remainder consisting of implicit opportunity costs. Energy subsidies are among the largest social expenditures in government budgets. Table 4.1 compares fuel subsidies from a recent IMF survey to public spending on health. Subsidies are 2 to 7.5 times as large as public spending on health in Bangladesh, Ecuador, Egypt, India, Morocco, Pakistan, Turkmenistan, Venezuela, and Yemen.

Removal of subsidies would be expected to increase economic efficiency and reduce GHG emissions over the long run.

Other sources point to additional countries with high subsidy-to-GDP ratios. Carey (2008), using IMF reports for 2006, lists Algeria (7.5 percent), Syria (12.2 percent), and Libya (15 percent). Indonesia's subsidies were $12 billion in 2005, and have since risen with fuel prices. As figure 2.3 shows, oil producers are prone to subsidize diesel fuel.

Emissions are markedly lower where countries have maintained high fuel prices through taxation.

Third, *removal of subsidies would generally be expected to increase*

economic efficiency and reduce GHG emissions over the long run. In the short run, people have limited options to react to price changes, especially where energy is rationed (for example, through load-shedding). Some analysts also assert that demand is insensitive to price in the long run (IEA 2007). But ample evidence shows that higher energy prices induce substantially lower demand, and, by extension, lower CO_2 emissions. Dahl and Roman (2004) reviewed 191 studies of energy demand since 1991. The studies found that a 10 percent increase in energy prices would be expected to reduce long-run demand by 7 percent, on average. Table 4.2 shows results for specific fuels.

Sterner (2007) compares gasoline demand across countries and shows that the decades-long price differentials among OECD countries have resulted in markedly lower demand in the countries that have maintained high fuel prices through taxation. In these areas, infrastructure and transport use patterns have evolved in a more energy-efficient manner. Sterner's results suggest that if the OECD had long ago harmonized prices at the level of the country with the highest tax (the United Kingdom), overall fuel consumption and emissions would be 36 percent lower. Had they coordinated at the lowest price (the United States), emissions would be 30 percent higher.

Subsidies to fossil fuels also boost CO_2 emissions by reducing the relative attractiveness of renewable energy. Subsidies to electricity similarly reduce the returns to investment in renewable sources.

At the global level, several studies show that removal of domestic subsidies leads not only to domestic gains but also to global improvements in welfare and reductions in GHG emissions. These studies trace the impacts of energy price changes through all interconnected markets. Anderson and McKibbin (2000) estimated that removal of coal subsidies (in both OECD and non-OECD countries) would reduce global CO_2 emissions by 8 percent from the business-as-

Table 4.1: Fuel Subsidies Compared with Health Expenditures

Region/country	Fuel subsidies (percent of GDP)			Ratio of fuel subsidies to public expenditures on health (%)
	2006	2007	2008	
Africa				
Angola	3.5	3.6	2.5	170
Burkina Faso	0.7	0.7	1.0	24
Cameroon	0.0	0.0	1.0	69
Cape Verde	1.9	0.0	0.0	0
Gabon	2.0	1.3	1.7	57
Mauritania	0.0	0.0	0.1	6
Mauritius	0.0	0.0	0.5	23
Nigeria	0.0	1.3	2.0	168
Senegal	0.6	0.3	1.3	77
Sudan	2.2	1.0	1.6	112
South Asia				
Bangladesh	1.8	2.9	3.0	362
India	1.6	1.2	2.0	213
Sri Lanka	0.8	0.3	0.4	21
East Asia and the Pacific				
Brunei Darussalam	1.0	1.0	1.0	63
Cambodia	1.5	2.1	2.6	168
Malaysia	1.3	1.4	2.8	149
Nepal	1.8	1.8	2.0	123
Pakistan	0.5	0.5	2.8	751
Europe and Central Asia				
Azerbaijan	2.2	0.9	0.2	21
Bosnia & Herzegovina	0.3	0.4	0.4	8
Belarus	4.4	4.9	7.4	148
Russian Federation	0.5	0.4	0.3	9
Turkmenistan	11.3	13.3	15.2	475
Ukraine	2.3	2.3	3.3	89
Middle East and North Africa				
Egypt, Arab Rep. of	8.3	7.0	8.4	362
Iraq	5.7	0.2	0.4	13
Jordan	3.6	4.5	2.5	53
Lebanon	0.1	0.1	0.2	5
Morocco	2.3	2.7	5.0	258
Oman	2.8	3.2	3.2	151
Tunisia	1.9	2.3	2.2	90
United Arab Emirates	2.9	2.7	2.5	134
Yemen, Republic of	8.3	9.4	11.6	544
Latin America and the Caribbean				
Argentina	1.1	1.7	0.0	0
Belize	0.1	0.4	0.4	16
Barbados	0.0	1.0	0.3	7
Costa Rica	0.0	0.0	0.5	9
Ecuador	5.6	6.4	8.7	410
El Salvador	0.0	1.4	2.0	53
Guatemala	0.0	0.0	0.4	19
Honduras	0.9	0.9	0.8	20
Mexico	0.0	1.6	2.1	72
Panama	0.0	0.5	0.5	10
Peru	0.0	0.2	1.0	47
St. Vincent and Grenadines	0.5	1.0	0.0	0
Trinidad and Tobago	0.8	0.8	0.8	35
Uruguay	0.0	0.4	0.4	11
Venezuela, Rep. Bol. de	4.6	5.9	7.7	363

Sources: Subsidies from IMF (2008). Health expenditure from World Development Indicators.

Note: Last column shows ratio of 2008 GDP share of fuel subsidies to 2005 GDP share of public expenditure on health.

Table 4.2: Sensitivity of Energy Demand to Price

Energy type	Long-run price elasticity of demand
Energy	−0.72
Industrial energy	−0.93
Electricity	−0.69
Electricity— industry	−0.32
Electricity—residential	−0.56
Coal	−0.60
Diesel	−0.67
Gasoline	−0.61
Natural gas—industry	−1.35
Natural gas —residential	−0.56

Source: Mean values from a metareview by Dahl and Roman 2004.

Studies suggest that subsidy removal leads to both domestic gains and global improvements in welfare and reductions in GHG emissions.

usual baseline, with little impact on GDP. IEA (1999) simulated the removal of energy subsidies in eight non-OECD countries and predicted a global reduction in CO_2 of 4.6 percent, while the countries concerned would improve GDP by an average of 0.7 percent. Saunders and Schneider (2000), using a lower baseline level of coal subsidies in the developing world, found a more modest 1.1 percent reduction in global emissions, but their model included more GHGs and international linkages than IEA (1999). Ivanic and Martin (2008) focused on the Middle East and North Africa, where energy subsidies are high. They found that removal of subsidies in the Region would boost welfare by $15.3 billion in the reforming countries and by $30.4 billion in the rest of the world, outside the Organization of Petroleum Exporting Countries (OPEC), though OPEC members outside the Middle East and North Africa would incur a $2.5 billion loss if supply was unchanged. Ivanic and Martin (2008) do not compute CO_2 impacts but note reductions of 7–30 percent in energy use in the reforming countries. There are, however, compensating increases in the rest of the world if oil conservation results in increased exports. In general, one would expect the greatest global impacts on CO_2 from the removal of electricity and gas subsidies.

There are considerable methodological challenges in quantifying subsidies.

The dearth of statistical information on subsidies is striking in view of their magnitude and economic and environmental importance. Aside from GTZ's invaluable biennial compilation of retail vehicle fuel prices, there is no comprehensive, public, and reasonably current source of comparative data on domestic energy prices. IEA discussed energy subsidies in its 1999 and 2006 *Energy Outlooks* but does not publish regular data on global subsidies. Its data on energy prices are mostly restricted to OECD members. The IMF sometimes discusses energy pricing and subsidies in its Article IV reports, and has just undertaken a selective rapid survey of subsidies—illustrating the feasibility of providing up-to-date information. Within the World Bank, the Latin America and Caribbean and Europe and Central Asia units have independently compiled useful summary tables of information on energy pricing for countries in their Regions, drawing in part on reports by regional associations, including the Latin American Energy Organization and Europe's Energy Regulators Regional Association. However, these compilations are not kept current.

There are considerable methodological challenges in quantifying subsidies. Many subsidies do not appear in government accounts. For instance, oil and gas producers and processors may be compelled to sell products at prices below alternative levels. Utilities may be required to sell electricity below the marginal cost of production, with indirect compensation. Or, perhaps more commonly, utility tariffs are set below long-run marginal cost, so that consumers are not faced with the cost of system expansion.

The recent rapid run-up in energy prices placed huge stress on existing subsidy systems. In some countries, this stress was unsupportable, and subsidies were scaled back. Elsewhere, rising prices translated directly into larger subsidies. This underlines the need for real-time monitoring of prices and subsidies.

Energy Subsidies and the Poor

Energy subsidies are often justified as protecting poor people, but the bulk of energy subsidies go to better-off consumers. Given the magnitude of subsidies, there is comparatively little information on their beneficiaries. However, the scattered information that is available shows that these subsidies are not well targeted. This is an almost automatic consequence of the relation between income and energy consumption. Most poor people in developing countries are not connected to the electric grid and do not own cars, so they get no direct benefit from fuel and gasoline subsidies. They do receive indirect benefits through lower prices of energy-intensive goods and services such as public transit. Nonetheless, a study by Coady and others (2006) found that even when such indirect benefits are considered, the bottom 40 percent of the population in Bolivia, Ghana, Jordan, Mali, and Sri Lanka received only 15 to 25 percent of fuel subsidies.

Appendix C presents a selection of information on the distribution of subsidies. The typical finding is that the bottom 40 percent of the income distribution receives 15–20 percent of the subsidies. Subsidies for liquefied petroleum gas and cooking gas are quite poorly targeted, because these items are consumed by better-off people. In Ecuador, the top quintile gets 17 percent of the cooking gas subsidy, the bottom quintile only 3 percent. In Bangladesh, the 4 percent of the population with gas access received 1.4 billion taka in subsidies.

Residential electricity tariffs are designed to subsidize poor people, but are often poorly targeted. Typically, tariffs increase with the quantity of electricity consumed or with the capacity of the connection. But even when rates rise with consumption, or differ by connection capacity, wealthier people derive large absolute benefits. In Indonesia, for instance, in 2005 the top decile received 44 percent more electricity subsidies than the bottom decile (World Bank 2006b). Komives and others (2006) reviewed 22 utilities (mostly in India) using quantity-based subsidies and found that none of them is progressive. Their targeting indicator is the ratio of the poor's share of subsidies to their share of the population (where the poor population is the bottom 40 percent of the income distribution).

This ratio ranges from .20 in Guatemala to 1.0 in Gujarat, with a median of 0.66. The poor targeting performance reflects low proportions of poor people connected to the grid and the persistence of subsidies for high consumers. However, Komives and others (2006) note three utilities that employ means-tested tariffs and achieve progressive targeting ratios of 1.2 to 1.5.

But even when richer people receive a larger share of the subsidy pie, poor people may derive a greater proportion of their income from those subsidies. For instance, though much of the subsidized kerosene intended for the poor is diverted to other uses, poor people nonetheless would be more burdened than the better-off by increases in kerosene prices.

The bulk of subsidies go to the better-off consumers.

For this reason, an understanding of the poverty and distributional impact of energy pricing reform would seem to be crucial for design of the reform and the monitoring of its impact. While long recognized, this was formalized only in 2004, with the Bank's adoption of the Operational Policy on Development Policy Lending (Operational Policy 8.60). This requires the Bank to determine whether a proposed Development Policy Loan (DPL) is likely to have significant adverse social impacts, particularly on poor and vulnerable populations; to summarize the state of knowledge on how to mitigate those impacts; and to fill gaps where necessary.

But some subsidies are important to poor people.

DPLs (and their predecessor, Structural Adjustment Loans) have frequently applied conditions related to energy pricing or subsidies. Figure 4.1 provides an indication of the application of conditionality over time, and includes general as well as sector-specific loans. It is an incomplete index because some conditions with pricing intent may be stated in an indirect manner.

Price reform policies should be guided by analysis and monitoring of poverty impacts.

Poverty and Social Impact Analyses (PSIAs) are one tool for fulfilling the Operational Policy 8.60 requirement.[3] First formalized in 2001, and piloted over 2002–03 by the Bank, the IMF, and

Figure 4.1: Conditionality Related to Petroleum Products

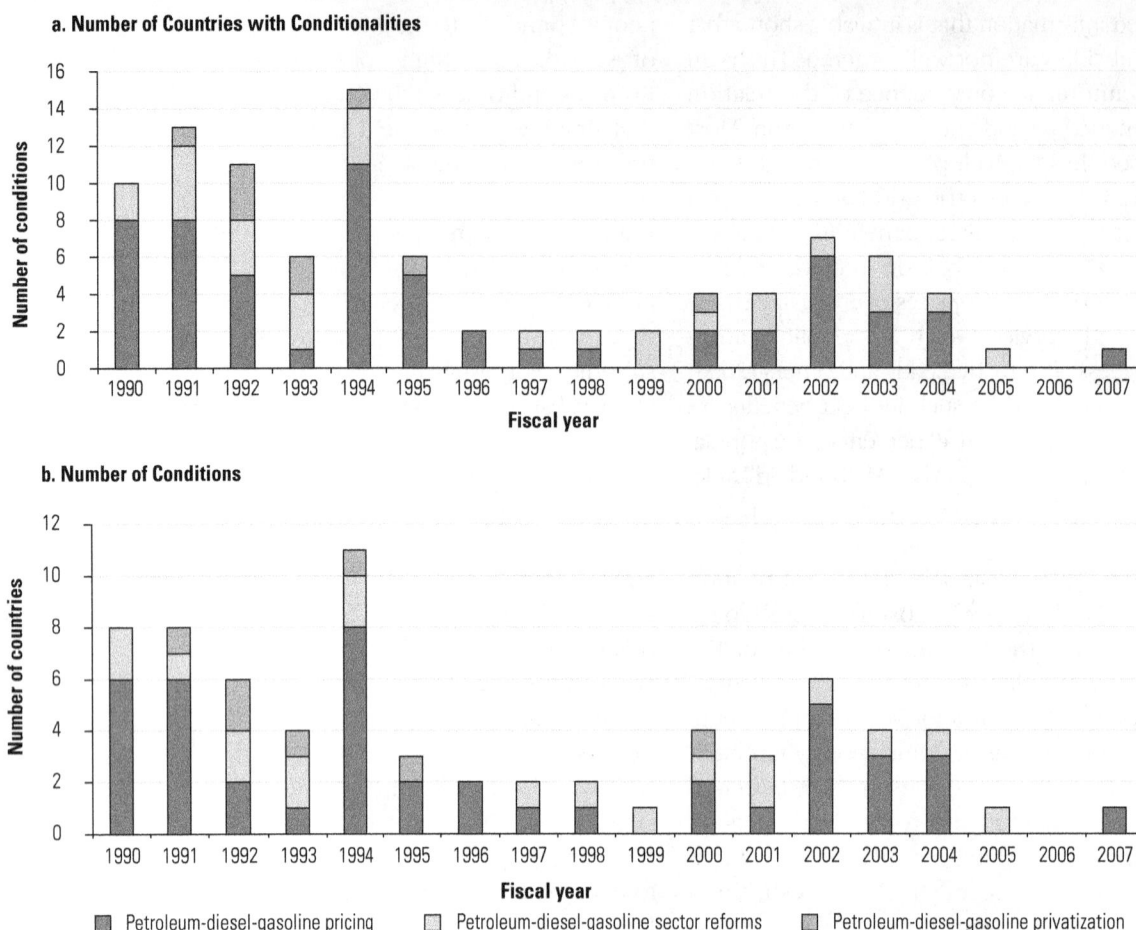

a. Number of Countries with Conditionalities

b. Number of Conditions

Legend: Petroleum-diesel-gasoline pricing · Petroleum-diesel-gasoline sector reforms · Petroleum-diesel-gasoline privatization

Source: IEG analysis.

others, they constitute a structured approach to assessing the distributional impacts of reform. They are not mandatory. A Good Practice Note (World Bank 2004a) encourages selectivity in undertaking PSIAs, prioritizing them where the distributional issues are most important, the policy options most precisely defined, and knowledge gaps the greatest.

PSIAs provide a structured approach to assessing the distributional impacts of reform, but they are not mandatory. IEG identified 19 completed PSIAs done by the Bank since 2001 that relate to pricing of electricity, heat, or fuels. All were done in the context of proposed price increases; a few were able to analyze retrospectively the effect of a previous price hike.

Most of the PSIAs attempted to document the current proportion of income devoted to energy expenditures by poor people, using this as a basis to assess the burden of price increases. Some also attempted to determine what coping strategies might be used by poor people to adjust to price changes. A few assessed specific policy alternatives for mitigating price impacts.

Accurate assessment of current budget shares and subsidy incidence requires good survey data. Some PSIAs used small purposive samples or focus groups, which cannot give reliable estimates of the magnitude of expenditure. Others commissioned custom surveys or were able to make use of existing high-quality, nationally representative survey data. However, survey data were often poorly suited to the task—for example, by failing to distinguish between billed and consumed electricity. Few of the PSIAs were able to undertake the sophisticated task of computing the indirect effect of price hikes on goods and services consumed by the poor.

PSIAs tended to make generic recommendations, such as advocating improved safety nets as compensation for price hikes or improved quality of utility service as a precondition for the political acceptability of raising tariffs.

IEG examined in more detail five Bank-supported PSIAs or PSIA-like analyses, together with an IMF-assisted PSIA for linkages to policy, Poverty Reduction Strategy Paper (PRSP), and loan outcomes. Outcomes were divergent. In two cases (see box 4.1 on Indonesia and Ghana), detailed PSIA analysis and specific recommendations appear to have shaped successful policies of price rises with compensation. In Yemen, which has some of the proportionally highest fuel subsidies, and where underpricing of fuel has encouraged unsustainable extraction of groundwater, long-standing policy dialogue drew on an ESMAP-sponsored study of household energy (not formally designated as a PSIA). The Bank recommended a combination of gradual price rises to allow time for adaptation, combined with improved targeting of the existing Social Welfare Fund. The government implemented enhanced social safety nets but raised diesel prices sharply—by 165 percent—rather than gradually. This triggered riots and 36 reported deaths, prompting a partial rollback of the price rise, which was still short of eliminating the subsidy.

In Egypt, a 2005 PSIA (World Bank 2005b) showed that energy subsidies are regressive. Although the poor and vulnerable receive a disproportionately small share of the energy subsidies, the PSIA concluded that removal of the energy subsidies would increase poverty. It recommended that the phasing out of the energy subsidies be coordinated with the development of a comprehensive safety net system. Fuel prices have subsequently risen, though they still fall short of world prices. A natural gas connections project, designed to shift consumers from highly subsidized liquefied petroleum gas to less-subsidized piped gas, refers to the PSIA, and the 2008 CAS describes ongoing policy dialogue in energy price reform.

In two cases, detailed PSIA analysis and recommendations seem to have shaped successful policies of price rises with compensation.

In two other cases, PSIA impacts were less clear. A 2005 PSIA of the Ghana electricity sector will be the subject of an in-depth IEG case study; for current purposes it suffices to note that the PSIA recommended raising tariffs while maintaining a relatively poorly targeted lifeline tariff (box 4.1). The government at first declined to change tariffs, but doubled them in 2007. In Bolivia, a 2004 study found that hydrocarbon subsidies were important to the poor, but leak substantially to non-poor households. The study is not cited in the Social Sectors Programmatic Structural Adjustment Credit or in the 2005 Poverty Assessment.

On balance, it appears that PSIAs or similar analyses have sometimes played a useful and substantial role in informing decisions on pricing reform. The availability of good survey data for use in the analyses appears to help, as does a substantial period of policy dialogue.

PSIAs have sometimes played a useful and substantial role in informing decisions on pricing reform.

PSIAs on energy often present generic recommendations for the use of targeted social safety nets. This is an area of increasing research and implementation at the Bank, following on the celebrated success of Mexico's conditional cash transfer program, PROGRESA-Oportunidades. Attention is being devoted to assessing the cost effectiveness and error rates of alternative methods of targeting (Castañeda and others 2005; Coady, Grosh, and Hoddinott 2004). Combinations of geographic targeting and means testing (or proxy means testing) offer favorable targeting performance and could be superior to fuel or electricity subsidies in this regard. As

---

Box 4.1: Ghana and Indonesia: Using Social Safety Nets to Protect the Poor from Fuel Price Rises

In Ghana, rising world prices led to increasing subsidies to the national oil refinery; these subsidies reached 2.2 percent of GDP in 2004. An IMF program pressed for reform of the sector. A government-commissioned PSIA, together with IMF-led research, analyzed the relative targeting effectiveness of a variety of specific compensatory mechanisms and recommended implementing educational or health benefits or means-tested transfers. These were predicted to be more effective in reaching the bottom quintiles of the population than the existing kerosene subsidies.

These analyses may have supported the 50 percent increase in fuel prices in February 2005 (which had been signaled the previous year) and probably helped to support announcement of a range of mitigatory measures, including elimination of fees for primary and junior secondary school, increased funding for primary health care in the poorest areas, investments in urban mass transit, and rural electrification. These measures remain in place. Prices for petroleum products have been linked to world markets. However, with the continued rise in oil prices, a gasoline tax (which funded some of the mitigatory measures) has been reduced.

Indonesia has long subsidized petroleum products, which has led to severe fiscal burdens. But eliminating the subsidies has been problematic. A 1998 price hike led to riots and is popularly thought to have contributed to the downfall of the Suharto government. Subsequent price hikes in 2000 and 2003 provoked protests. Attempts to compensate poor people met with little success (Bacon and Kojima 2006). In 2005, the government confronted subsidies reaching 7 percent of GDP. The government drew on analytic work from many sources, including a PSIA. The PSIA showed that the fuel subsidies were regressive and that past mechanisms to compensate the poor for price hikes had been ineffective. It suggested a geographically targeted cash transfer mechanism. The government subsequently undertook two large price hikes of fuels, including kerosene, in tandem with a means-tested, unconditional cash transfer system. Thanks to the Indonesian statistical bureau's well-developed household survey system, the government was able to develop and implement the targeting and transfer system within a couple of months. The price of diesel fuel doubled and that of kerosene nearly tripled; but monthly cash payments of $10 were distributed to each of 19.2 million households for a year. Subsequent simulation analysis suggests that, even if substantial mistargeting is assumed, the bottom four deciles of the population gained during the period of transfer (World Bank 2006d). However, the continued rise in oil prices has again boosted subsidies. In May 2008, fuel prices were again raised, and the cash transfer program continues.

Sources: Azeem (2005); Bacon and Kojima (2006); Coady and others (2006); World Bank (2005e).

---

countries face a combination of high energy and high food prices (often in conjunction with food subsidies), interest in unified social protection systems grows.

Experience in the Transition Economies

In the transition countries, the Bank supported reforms that made rapid pricing adjustments and, in most cases, reduced emissions.

Although their experience was historically singular, the transition economies of Eastern Europe and the former Soviet Union provide interesting examples of massive and rapid adjustments in energy pricing. These adjustments have been accompanied in most cases by sharp reductions in emissions intensity. They certainly reflect the huge structural changes in the economies, but the structural changes themselves were entangled with changes in energy pricing.

In Ukraine, a combination of fiscal stress, government ownership, and cross-sectoral coordination of reforms, DPLs, and analytic work has facilitated price adjustments and a reduction in emissions intensity (box 4.2).

In Romania, the 1995 CAS aimed at pricing reform in the energy sector and was supported by investment and adjustment lending. By 2001, energy sector subsidies were estimated by the IMF (Cossé 2003) at 5.2 percent of GDP, including 3.3 percent in off-budget transfers to industrial users. However, IEG (2005) found that significant reforms were spurred mainly by conditionality attached to EU accession, IMF standbys, and the World Bank Programmatic Structural Adjustment Loan 2 in 2002–03. Electricity tariffs increased (in 2001 prices) from

---

Box 4.2: Ukraine: Gradual Energy Policy Reform and Decreasing Emissions Intensity

Ukraine began the transition period of the 1990s with one of the world's most energy-inefficient and emissions-intensive economies. Energy prices were far below economic levels and collection rates were low. Restructuring of the energy sector began in 1994 with some early successes, including shutdown of uneconomic coal mines, but subsequently faltered. A Bank-supported Electricity Market Development Project (fiscal 1997), failed because of a premature approach to privatization. Meanwhile, the economy was suffering a severe contraction from pretransition levels, exacerbated in 1998 by the regional crisis triggered by a sharp rise in the prices of imported energy. The crisis forced the government to introduce stabilization measures and to agree with the IMF on a standby arrangement and with the Bank on a financial sector adjustment loan operation in late 1998.

During the current decade, energy reforms have accelerated and have been supported by the Bank through analytical inputs and policy advice as well as investment. The Bank's energy policy engagement in the 2000s has included three policy-based operations (Program Adjustment Loan I and II and DPL I) and extensive analytical work, including reviews of energy sector reform options, electricity, mining, and gas markets. The supervisory committee established for each policy-based operation included not only the line ministers, but also other key cabinet members, including the minister of finance, and the central bank governor. There were multiple working groups on each of the main themes in the Bank's

policy-based loans, including energy. The supervisory committee and working groups helped improve information flows across government agencies and provided a forum to design strategic decisions, as well as for monitoring their implementation. The policy-based operations were complemented by a set of sectoral loans that focused on energy efficiency.

Pressure for reform was driven in part by a large quasi-fiscal deficit—6 percent of GDP in 2003—which had resulted from a sharp increase in the imported energy prices that had been heavily subsidized until then. Sector reforms resulted in a tremendous increase in cash collection rates, rising from 8 percent in 1999 to 98 percent in 2005, thus establishing an effective demand price for energy consumption. State-regulated energy tariffs (electricity, gas, and coal) were increased between 25 and 50 percent during the 2002–07 period, with additional pressure coming from a hike in the price of imported gas. Economy-wide energy efficiency increased, and CO_2 emissions/GDP dropped from 5.6 tons/\$ in 1998 to 3.9 in 2005.

The impact of the increase in energy payments on the population was partially cushioned by vigorous economic growth and a corresponding decrease in poverty during this period. In addition, Ukraine already had social support mechanisms in place to protect the most vulnerable. To further protect the poorest against the rise in energy tariffs, the government also introduced a graduated (lifeline) tariff in November 2006 to those whose utility payments exceeded 20 percent of their income.

Sources: IEG 2008c; IEG staff.

3.8 to 7.9 cents per kWh from 2001 to 2005, and collection rates went from 49 to 99 percent. However, safety net features of the Programmatic Structural Adjustment Loan 2 were focused on unemployment and pensions rather than on energy prices.

Georgian power reform followed an uneven path (IEG 2008e; Lampietti, Banerjee, and Branczik 2007). Bank-supported reform commenced in 1995. An independent regulator was established, and Tbilisi's power system was privatized. There was a dramatic hike in collection rates across all income groups, spurred by an aggressive system of re-metering. Tariffs also increased steadily. In-kind transfers intended to help the poor and

pensioners reached 37 percent of the bottom quintile, but went to 24 percent of the top quintile as well. Overall, the average share of household expenditure going to energy stayed constant, as did energy consumption. Reform rates faltered, but revived with the Rose Revolution of 2003, and subsidies for electricity fell from 6 percent of GDP to zero. Concerns remain about the independence of the regulator.

Overall, the transition experience shows that progress in rationalizing energy prices is possible. Fiscal stress and the prospect of EU access have served as incentives to undertake difficult reforms. Dramatic improvements in collections have taken place, in part due to

Fiscal stress and the prospect of EU accession helped spur the reforms.

privatization and metering. Coordination across sectors is an issue; increases in heating tariffs shifted consumers to subsidized gas, threatening the viability of the heating systems. The result of these changes has been substantial reductions in emissions intensity of these economies. But the impact on welfare of the poor is not well understood and needs to be better tracked (Lampietti, Banerjee, and Branczik 2007).

Bank Engagement with the Large Subsidizers

Roughly speaking, the largest impact on global emissions might be expected from the largest subsidies in absolute magnitude. This section looks at the Bank's engagement on these issues with the largest subsidizers among its borrowers, based on the IEA (2007) list of large, non-OECD subsidizers of 2005. The list is augmented with Mexico, a large electricity subsidizer and member of the OECD. (See appendix A.) The analysis looks at the role of Public Expenditure Reviews (PERs) in identifying and drawing attention to subsidies, and of CASs in prioritizing action. PERs, introduced around 2000, are of interest because one might expect this to be an apt tool for detecting and diagnosing subsidy issues.

Returning to appendix A, there are eight countries without PERs. These include Egypt, Iran, Kazakhstan, and Venezuela, where implicit subsidies are large in both absolute terms and as a proportion of GDP. Some countries without PERs, such as China, India, and Vietnam, nonetheless include detailed treatment of subsidies and pricing in their CASs or Country Partnership Strategies (CPSs). Among the countries with PERs, energy subsides receive no mention or only a perfunctory mention in the documents for Argentina, Malaysia, Nigeria, and Pakistan, although the Pakistan CASs devote significant attention to pricing issues. In the remaining countries, PER treatment of subsidies includes detailed analyses and recommendations.

Outcomes of Bank engagement with the largest subsidizers have varied greatly.

Outcomes of engagement vary greatly among countries. China, Indonesia, Russia, and Ukraine stand out as examples of long-term engagement associated with bold action by governments to rationalize prices. India, Mexico, and Pakistan, in contrast, present a sequence of CASs that repeat the same set of concerns, but with relatively modest apparent impact. In Vietnam, there has been slow progress on prices, with little or none in Argentina, Kazakhstan, and Nigeria. In Egypt, analytic work on pricing may have contributed to recent natural gas and fuel pricing decisions. The two biggest diesel and gasoline subsidizers are Iran, where there has been some analytic work, and Venezuela, where there has been no engagement on this issue.

Russia presents a complex record of Bank involvement and price reform. There were 10 loans with primary fuel pricing or subsidy reduction objectives during 1993–99. The three coal sector loans succeeded in drastically reducing subsidies to loss-making coal mines—a difficult task that has challenged many countries. The Russian effort combined extensive, effective safety nets and job creation programs for the affected communities and improved, transparent systems for managing and winding down the subsidies. At the same time, a series of loans directed at oil and gas market reform were mostly unsuccessful. One loan with gas pricing objectives, for instance, focused on gas users rather than the gas supplier and failed to achieve its objective. Since 1999, however, gasoline and diesel prices have increased to world market levels. Gas prices have increased but remain below netback (export parity) levels.

Energy Loans and Pricing

This section considers the global experience with pricing-oriented loans in the energy sector. Figure 4.2 tallies experience with such loans, distinguishing between those concerned with primary energy (petroleum products and gas) and those involving electric power. It shows a decline in loans dealing with primary fuel pricing. Few countries have had more than one such loan. There is, however, an apparent post-2000 increase in loans dealing with power pricing.

Box 4.3: Egypt: Policy Dialogue and Pricing Reform

Egypt has long maintained energy subsidies. After bitter disagreements with the Bank on tariffs and financial management, the fiscal 1992 Kureimat Power Project was closed in fiscal 1994 and $199 million of the $220 million loan was canceled. This effectively ended the Bank's lending role in the power sector in Egypt until 2006, even though it did provide advisory services during this period.

By 2004, rising international energy prices had boosted the cost of Egypt's energy subsidy policies to more than 8 percent of GDP and prompted renewed attention to the country's social safety net policies. A retreat at Luxor in February 2005, led by the prime minister and the World Bank president, included most of the Cabinet and brought senior officials from Mexico and Brazil, who presented their experiences with safety nets. The retreat was followed by a joint study entitled *Egypt—Toward a More Effective Social Policy: Subsidies and Social Safety Net* in December 2005 (World Bank 2005b). The report demonstrated clearly that "energy subsidies distort economic decisions and benefit the rich more than the poor." The Bank also provided a set of reform options for the energy and food subsidies.

Underpricing of energy has been a problem for the country since the early 1990s, and remains so today, though recent years have seen an increase in tariffs. Prices of electricity were adjusted in October 2004 (from an average of 12.8 Pt/kWh, 2.2 cents, to 14.06 Pt/kWh, 2.4 cents) for the first time since 1992. Electricity prices were increased at an average rate of 8.6 percent in 2004, by 5 per-

cent in 2005, and by 7.5 percent in 2006. A further 5 percent annual increase is planned for the next five years. The prices of gasoline, diesel, fuel oil, and natural gas increased in 2004, 2006, and 2008. The Ministry of Finance started to record energy subsidies in the budget in 2005/06 to increase transparency. In August 2007, the government announced plans to eliminate gas and electricity subsidies for energy-intensive industries over the coming three years to help reduce budget deficits.

These structural reforms have been homegrown, as a result of government initiative, and have drawn on Bank advisory services. The Bank had traditionally engaged on energy issues with the Ministry of Petroleum and the Ministry of Electricity and Energy, but began to work with the Ministry of Finance in 2005. The Bank arranged an international conference in September 2005 on DSM and energy efficiency. In 2006, two studies were delivered, one on the Economic Costs of Gas in Egypt and another on the Load Management Program and Time Use of Tariffs. An interministerial steering group, which includes the minister of finance, the minister of investment (economy), the minister of petroleum, the minister of electricity and energy, the minister of social solidarity, and the minister of industry and trade, was established in late 2007 for working with the Bank on the energy pricing strategy. The overall objective of this study is to formulate an energy pricing strategy that ensures energy price levels are reflective of the underlying economic costs. The study is scheduled for completion in 2009.

Source: IEG staff.

Figure 4.3 maps countries where there has been project-level engagement in electricity pricing. India and China stand out as areas of significant engagement.

Table 4.3 summarizes the impact of loans oriented toward electricity pricing for the subset with detailed IEG audits. This is not a random sample, but it provides a wide set of experience for which outcome information is available. The overall message is one of general success in transition economies, including Armenia, Bosnia and Herzegovina, Bulgaria, China, and Georgia, sometimes through structural adjustment credits and sometimes through investment loans. The well-documented Armenian case

(through an ex-post PSIA; Lampietti and others 2007) is noteworthy for its inclusion of an alternative social safety net, although that protection failed to reach half the poor. In other countries, the experience was mixed or unsustained.

Additional evidence comes from the experience in India. The World Bank has completed five state-level power sector restructuring projects in India, in Orissa (fiscal 1996), Haryana (fiscal 1998), Andhra Pradesh (fiscal 1999), Uttar Pradesh (fiscal 2000), and Rajasthan (fiscal 2001). IEG rated the outcomes in Orissa, Haryana, and

Pricing-oriented lending for primary fuels has been declining, while that for power has increased since 2000.

Lending for power-price reforms has generally been successful in transition economies.

51

Figure 4.2: Trends in Energy Sector Loans with Pricing Goals

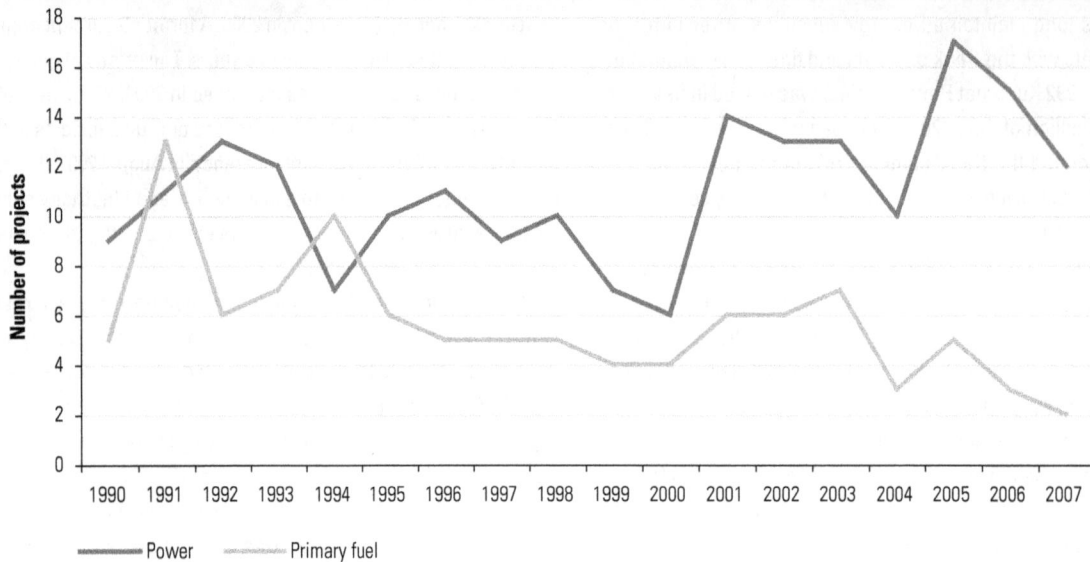

Source: IEG tabulation.

Rajasthan as moderately unsatisfactory, and that in Uttar Pradesh unsatisfactory. Only the Andhra Pradesh project was rated satisfactory. All sought to privatize distribution and establish or strengthen independent regulatory commissions with the aim of reforming tariffs, and thereby boosting the sector's financial health. All the projects fell short of the desired tariff objectives, in part due to a premature emphasis on privatization. This goal of removing the subsidies for farmers proved the subsidies to be too deep-seated an issue to be addressed through power sector reform. However, some of the states have since made progress in increasing collection rates and improving financial sustainability.

Attempts to reduce power subsidies to agriculture failed.

We identified 107 loans with goals related to primary fuel pricing since 1990; 66 of these were DPLs or adjustment loans. We were able to assess the outcomes of 50 post-1995 loans, most of which were DPLs. Thirty of these reported achievement of pricing goals, though sustainability was not assessed.

Conclusion

Price reform in energy is more urgent than ever, given the run-up in international market prices. In principle, adjustment to higher and more volatile energy prices could yield fiscal dividends and long-run reductions in the level of GHG emissions. Also in principle, reallocation of the savings from lower subsidies and lower energy use could benefit poor people and society at large. But in all societies the adjustment costs are large, especially for those who have benefited most from low prices.

Over the past 18 years, the World Bank has frequently supported energy subsidy removal or price rationalization through both investment and policy lending. Price reform goals have often been at least partially achieved, especially in transition economies. In many transition economies, price reform has accompanied absolute declines in emissions per capita and per unit of GDP, while incomes have risen. But in many cases, prices remain below the long-run marginal cost.

Figure 4.3: Distribution of World Bank Lending Related to Electricity Power Pricing Policy, 1996–2007

IBRD 36600

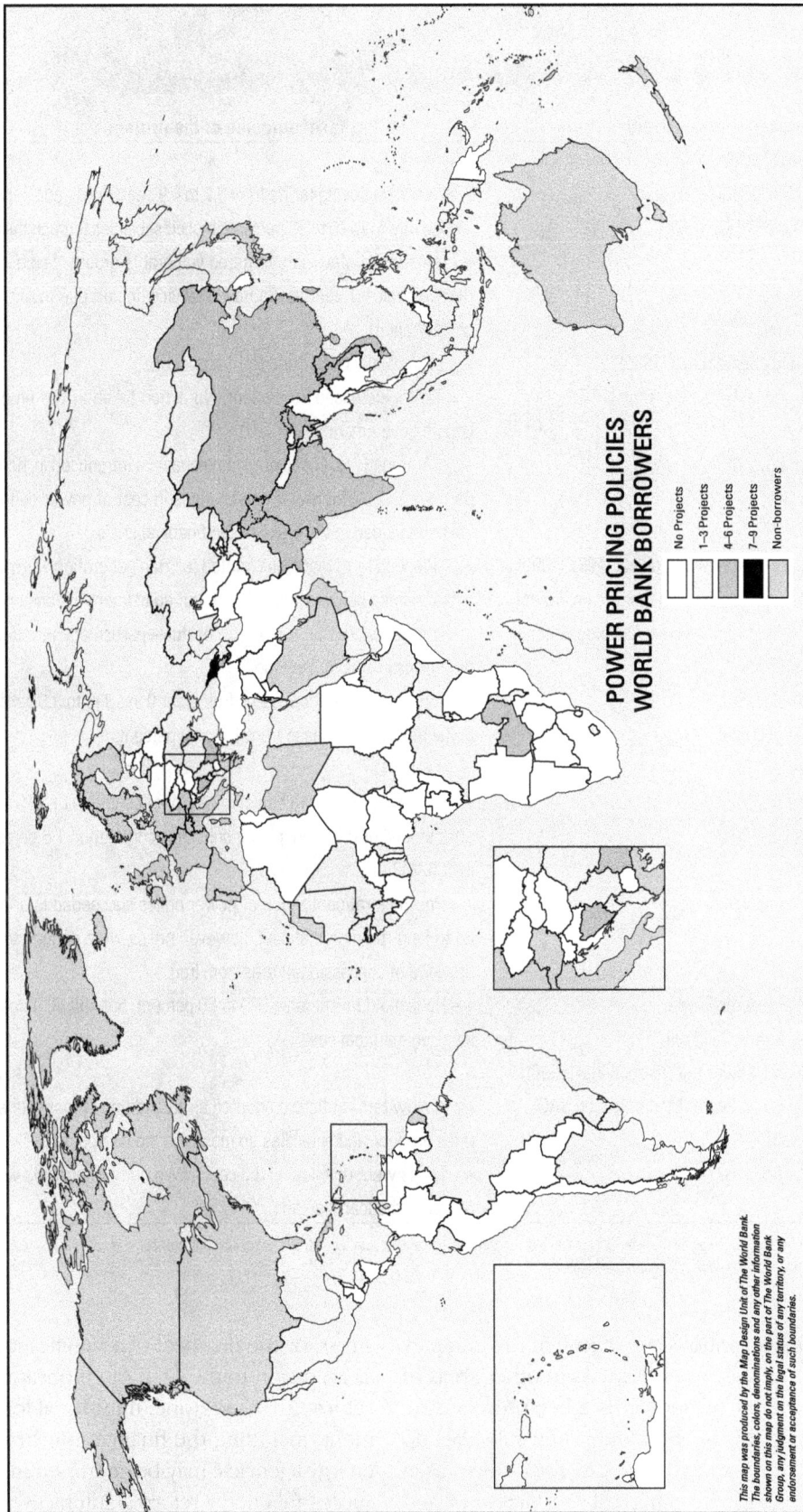

POWER PRICING POLICIES
WORLD BANK BORROWERS

No Projects
1–3 Projects
4–6 Projects
7–9 Projects
Non-borrowers

This map was produced by the Map Design Unit of The World Bank.
The boundaries, colors, denominations and any other information
shown on this map do not imply, on the part of The World Bank
Group, any judgment on the legal status of any territory, or any
endorsement or acceptance of such boundaries.

MARCH 2009

53

Table 4.3: Outcomes of Loans with Electricity Tariff Goals

Country	Project (year approved)	Tariff outcome of the project
Armenia	SAC I (1996) SAC II (1998) SAC III (1999)	+ + Succeeded in raising tariffs from 0.2 to 4.9 cents/kWh, and household collection rate from 10 to 88 percent; implicit subsidies through the water sector remained. Safety nets targeted vulnerable groups. Established quantitative goals for utilities. Improved service (in part due to restart of large nuclear plant).
Côte d'Ivoire	Energy Sector Loan (1990)	– Economically unjustified tariff reduction.
Laos	Provincial Grid Integration (1993)	+ + Tariffs increased 70 percent. No action taken at the time to reduce unpaid government bills.
Indonesia	Suralaya Thermal Power (1992)	+ / – Automatic tariff adjustment mechanism introduced in 1994 was only partially successful in tracking changes in cost of power generation and was abandoned in the wake of the financial crisis.
China	Ertan Hydropower Projects (1991, 1995) SN Sichuan Power Transmission Project (1995); Zhejiang Power Development Project	+ / – Planned tariff increases related to Ertan generation were inadequate and remained below marginal cost, but adjustment of consumer tariffs in the other two projects was successful—particularly introduction of time-of-day rates in Zhejiang.
Georgia	SAC I (1996) SAC II (1998) SAC III (1999)	+ + Tariffs raised in three steps from near 0 in 1995 to 3.5 cents in 1997. Collections rate increase from 10 to 65 percent.
Pakistan	SAC I (2001) SAC II (2004)	+ / – Power tariffs were adjusted as a prior condition of SAC I, but thereafter stalled or reversed; power subsidies constituted 1.6 percent of GDP in 2002/03.
Jordan	ESL (1994)	+ + One-time rationalization of power prices succeeded in bringing them up to long-run marginal cost; however, prices were pegged to oil prices, provided at concessional rates from Iraq.
Bosnia and Herzegovina	EMG Electric Power Reconstruction (1997) Electric Power Reconstruction II (1998)	+ + Household tariffs raised 20 to 60 percent, but still 40 percent below long-run marginal cost.
Honduras	HN Public Sector Modernization SAC (1996)	+ / – New tariff structure adopted as a condition of loan, but average rates are low, and subsidies go mostly to the non-poor.
Bulgaria	REHAB (1997)	+ + Tariffs were doubled, to 3.3 cents/kWh and adjustments were continued after the loan's ending.

Note: + + = general tariff increase of more than 10 percent; + = tariff increase of an unspecified percentage less than 10 percent or covering only some residential consumers; – = tariffs decreased during and/or after the project; + / – = mixed or unsustained results.

While generalizations are difficult in this complex area, some lessons emerge. As in other areas of reform, client ownership is a key prerequisite. Engagement is often lacking when subsidies do not cause immediate fiscal stress, as in the case of implicit subsidies to oil and gas in net exporters. Conversely, fiscal stress, or the prospect of a significant gain (such as accession to the EU), can motivate interest in reform. Cross-sectoral, ministerial-level involvement, including the finance ministry as well as energy agencies, may be an important feature of successful energy reforms. Interactive and client-responsive policy dialogue over an extended

Price reform goals have often been achieved.

period, supported by strong analytic work, is another recurrent theme, though it does not guarantee results. Overcoming vested interests, especially highly subsidized agricultural users, has been difficult, even in the presence of such work.

In at least two cases—Ghana and Indonesia—the availability of pre-existing, good-quality household survey information on welfare and on energy consumption helped in assessing the impacts of price reform and in the design of programs that mitigated adverse impacts on poor people. But there has been a lack of the systematic monitoring of energy expenditure and usage that would permit real-time and long-term assessment of welfare and emissions impacts.

These and other examples point to growing interest in scrapping energy subsidies in favor of more efficient and integrated social protection systems. Systems using proxy means testing and geographic targeting could be more effective in helping poor people and could free up resources for investment in energy efficiency. But little effort has been made to use the introduction of energy efficiency as an adjustment vehicle for higher tariffs.

But the Bank has found it difficult to engage in price reforms in petroleum- or gas-producing countries not under fiscal stress.

Chapter 5

Evaluation Highlights

- The Bank's energy efficiency projects have had high domestic and global returns.
- Five percent of the value of the Bank's energy lending has been for end-user efficiency and district heating projects.
- Only a handful of projects have effectively supported efficiency policy, though there is institutional innovation in efficiency finance.
- Bank and borrower incentives favor supply over efficiency.
- Modest GEF and trust fund finance has supported long-term policy engagement on efficiency issues.

Solar energy is used to light village shop, Sri Lanka. Photo by Dominic Sansoni, courtesy of the World Bank Photo Library.

Efficiency Policies

E nhanced energy efficiency is seen by many as the largest single win-win opportunity to reduce emissions. Ultimately, people care less about energy than they do about the services it provides. And much energy is simply wasted.

The famous satellite photo of the world's urban lights is a graphic illustration of energy being cast uselessly into space, where, as light pollution, it spoils enjoyment of the nighttime sky. Coal plants throw off two-thirds of the energy they burn—energy that in principle could be captured for heating or industrial purposes. Energy costs money, so efficiency offers the prospect of reducing emissions at negative cost.

According to the McKinsey Global Institute (Bressand and others 2007), growth in global energy demand could be halved through investments with financial rates of return over 10 percent. IEA (2007) identifies increased energy efficiency as a crucial and cost-effective component of a global energy strategy over the coming decades. In the electricity sector, it estimates that non-OECD countries would save $3 in supply investment for each $1 in demand-side efficiency investment. (Fuel savings would be an additional payoff.) Across all energy sectors, the IEA estimates that there is scope for nearly a trillion dollars of efficiency investments in non-OECD countries over the next 25 years.

These analyses are not novel. As noted earlier, the World Bank's 1993 energy policy pointed to increased energy efficiency as an important area for attention. And global energy efficiency (measured as GDP$ per unit of energy consumed) increased over 1990–2005, particularly in China, India, and Russia (IEA 2008b). But end-user energy efficiency appears to be elusive: for at least 20 years, energy experts have pointed to high-return opportunities that have been missed.

Overcoming the Barriers to Energy Efficiency

The persistence of high-return opportunities for end-user efficiency seems paradoxical. There is a standard set of explanations of the market and policy failures that result in barriers to the pursuit of these opportunities. These include:

- *Information failures*—Firms and households cannot gauge the potential for energy savings, are unwilling to pay for an energy audit that may fail to identify savings, and doubt whether efficiency investments will be as profitable as advertised.

Standard explanations for the barriers to energy efficiency include market and policy failures.

- **Financial market failures**—Banks do not know how to appraise loans for efficiency improvements, or perceive this to be an unusually risky business.
- **Attention failures and transactions costs**—Where energy costs are a small part of overall expenditures, other issues and opportunities may command decision makers' attention. For instance, neither consumers nor manufacturers pay much attention to standby power demands of appliances, which may be only a few watts each, but the aggregate national burden of these appliances on the power system could be large.
- **Split incentives**—If buyers or renters cannot gauge the energy costs of buildings, builders may have no incentive to invest in costly but energy-saving construction methods.
- **Other incentive or regulatory failures**—Unmetered heat consumers may lack the incentive or means to adjust temperatures; public agencies may be barred from considering life-cycle costs in procurement decisions; and utility rate-setting procedures may reward inefficiency.
- **Underpricing of energy,** so that users lack incentives to conserve it.

There is a standard set of remedies for these failures. These can be roughly categorized along two dimensions: supply versus demand and investment versus policy. (See table 5.1.) Supply-side interventions concentrate on the generation or transmission of electricity, and are often components of sector reform efforts. Demand-side interventions focus on the behavior and technology of energy users, and are therefore more diffuse and varied.

The investment-versus-policy distinction is crucial to the discussion in this report. It roughly corresponds to retail-versus-wholesale intervention. Investment projects without strong policy components intervene directly to install or fund efficiency measures. On the supply side, these include measures such as improved boilers in power or district heating plants or transformers in electric distribution, so that more electricity is delivered per unit of fuel burned. On the demand side, analogous projects fund installation of insulation, efficient lights, or improved electric motors.

In contrast, policy interventions seek to remove the barriers that inhibit firms and households from pursuing these investments themselves. Supply-side policies might encourage these investments through incentive changes—for instance, through corporatization of a utility. Demand-side policies include standard setting or certification systems that establish efficiency

Table 5.1: A Typology of Efficiency Interventions

	Investments	Policies
Supply side	District heating renovation Combined heat and power Coal boiler renovation Improved transformers	Power sector restructuring
Demand side	Distribution of compact fluorescent light bulbs (CFLs) Building retrofits District heating renovation Funding for energy financial intermediaries including energy service companies	Utility DSM Appliance and building standards and certification Public procurement policies Capacity building and promotion of energy service companies

Source: IEG.

requirements for appliances or enable consumers to reliably distinguish differences in efficiency. The IEA has recently published a comprehensive set of public policy recommendations related to energy efficiency (IEA 2008a).

This Phase I report concentrates primarily on policy-level interventions; analysis of investments will be presented in Phase II. That report will look at the impact of policy changes on the diffusion of efficiency technologies. However, there is not a sharp distinction between policy and project. At the intersection is an area of increasing Bank Group activity: efficiency finance and promotion of energy service companies (ESCOs). At the core of these projects is the goal of introducing new mechanisms to identify and finance retrofits that make equipment or buildings more energy efficient. Sometimes there are also public policy elements in promoting capacity and demand for these services, and in some cases public funds are provided. Similarly, public policies may be used to promote the diffusion of new technologies, such as compact fluorescent light bulbs. The second phase of this evaluation will look in more detail

at the Bank Group's efforts at ESCO promotion, which constitute a significant and interesting segment of the efficiency portfolio.

The Efficiency Portfolio

Figure 5.1 draws on data and categorizations presented in the Bank Group's reports on renewable energy and energy efficiency.[1] It shows the total value of energy-related lending at the World Bank, and the proportion devoted specifically to energy-efficiency components, including both supply- and demand-side efficiency.[2] Over the period 1991–2007, about 5 percent of the value of the World Bank's $48.7 billion energy investments were devoted to energy efficiency. (These figures exclude IFC and MIGA.) The proportion has oscillated widely from year to year. This 5 percent of value was concentrated in about one-tenth of all projects. Among high-emitting countries (table 3.2), this proportion was 6.7 percent for countries with high-level CAS/Country Partnership Strategy goals related to efficiency, and 3.9 percent for the remainder. A broader count, including transmission and generation rehabilitation projects deemed by the World Bank Energy Anchor to

Figure 5.1: Energy-Efficiency Investments

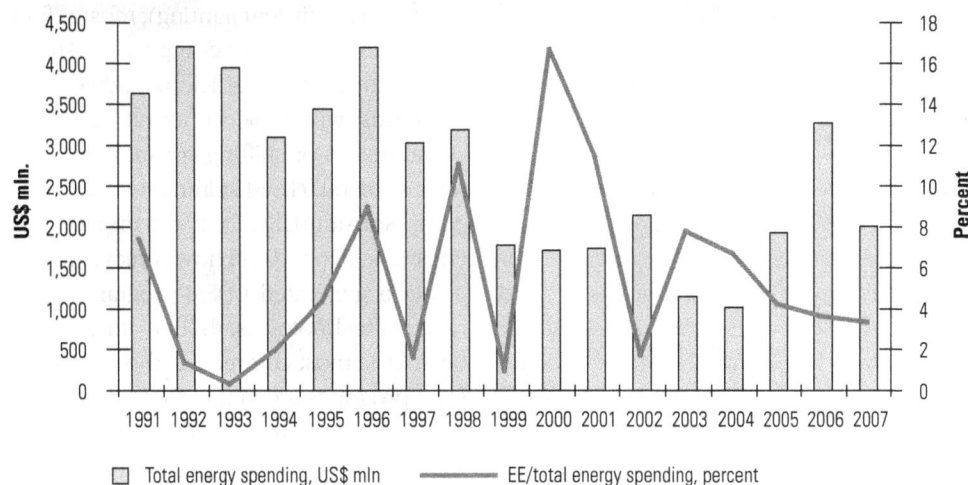

Total energy spending, US$ mln — EE/total energy spending, percent

Sources: Total energy spending: IEG computation based on energy-designated proportion of project commitments. Efficiency component spending: *World Bank Progress on Renewable Energy and Energy Efficiency* reports 2004, 2005, 2006, 2007 (World Bank 2005d, 2005b, 2006a, 2007b). Scope of coverage was more precisely defined for 2006 and 2007. Includes Bank-executed medium-size and large GEF projects. Excludes IFC and MIGA.

increase efficiency or reduce emissions, exceeds one in five for 1998–2007.

While this evaluation is concerned primarily with the World Bank, it is important to note that IFC investments in energy efficiency have grown sharply since the Bonn Commitment of 2004 and now exceed those of the World Bank. IFC commitments for energy efficiency were $94 million over 1991–2004, but jumped to $621 million over 2005–07.

The World Bank has also undertaken analytical and advisory work in this area. One line of action is in support of Green Investment Schemes in Europe and Central Asia. Some Eastern European and former Soviet Union nations, parties to the Kyoto Protocol, have an excess supply of assigned amount units (carbon allowances). Some would-be purchasers of these units seek assurances that sale revenues will be used for emissions reduction or other environmental services. The World Bank has helped some of these countries explore how such revenues could be used to support energy efficiency. For example, in Bulgaria, a country where the World Bank has supported energy efficiency, Green Investment Scheme potential was identified in cogeneration and energy efficiency.

Although energy-efficiency investments consti-tute a small proportion of all investments in energy, they offer economic returns that are as good or better than those of other investments in the sector. (See box 5.1.)

IEG coded energy-efficiency projects for policy content related to end-user efficiency, including regulatory provisions for DSM; appliance or building standards and certification; and research, demonstration, or planning of energy efficiency during 1996–2007. This policy coding excluded projects confined to engineering activi-ties. A total of just 34 projects met these criteria (see appendix C), about one-third in the last third of this period.

This is a heterogeneous list, including many projects—such as a small study—with only minor policy and regulatory components. Others, such as the Morocco DPL, aim at economy-wide impacts. The list contains nine projects that deal with standards and codes, and nine that deal with utility-based DSM; these topics are discussed at greater length below. There are at least seven projects on the list that deal with ESCOs or efficiency finance.

Energy-efficiency staff in the World Bank are relatively few: about 22 staff members work a substantial portion of their time in this area. A few additional staff are slated for recruitment under the Energy Efficiency for Sustainable Development Scale-Up Strategy and Action Plan, a program announced in 2007.

The remainder of this section examines key areas of the portfolio for policy content, including DSM, appliance and building codes and standards, district heating, and public buildings. We touch only briefly on supply-side efficiency, which is difficult to disentangle from more general energy investments.

Demand-Side Management

DSM programs are traditionally run by utilities to encourage customers to reduce and shift their energy use. Program design can vary. Some examples of programs include marketing efforts to encourage customers to adopt new technolo-gies (such as efficient lighting), rebates for certain types of energy-efficiency equipment, free energy audits, and energy-efficiency advice. Typically, programs will look either at reducing total consumption or shifting consumption to reduce peak demand. (Load shifting reduces investment costs substantially, but its impact on GHG emissions depends on how peak versus baseload power is generated.) DSM programs can cover a range of technologies, though efficient lighting is one of the most common of those covered. Most DSM programs are in the power sector, although there are programs for other types of energy as well, including natural gas, district heating, and fuel oil. While DSM can be in a utility's financial interest (particularly for load shifting), sustaining DSM in the long term requires supportive regulations.

Limited evidence suggests that energy-efficiency projects offer attractive domestic economic rates of return (ERRs) that are greatly enhanced when the global benefits of emissions reduction are factored in. Over 2000–05, six Bank projects supporting energy efficiency closed, all of them involving supply-side or demand-side improvements to district heating systems. ERRs ranged from 22 to 44 percent, not including environmental benefits. Including those benefits, which comprised reductions in local air pollution as well as CO_2, the ERRs of projects and sub-projects ranged from 27 to 289 percent. This compares to an un-weighted mean of 22.4 percent for the 45 nonrenewable power sector projects that closed over the same period.

Transmission projects are not included in the tally of efficiency projects above, but reductions in transmission losses offer potentially high efficiency and GHG mitigation gains, require large investments, and may include a role for public policy because of the regulated or monopoly nature of most transmission systems.

Experience has been variable. Projects in Albania and Uganda failed to achieve their objectives, with continued nontechnical losses (that is, theft) that could not be distinguished from technical losses. In Macedonia, Serbia, and Zambia, ERRs were 111, 18, and 35 percent. In India, significant reductions in transmission losses were achieved in Rajasthan and Orissa. The rates of return depend on the true economic valuation of electricity, which is above the tariff level but difficult to estimate. Depending on this value, the ERR of the Rajasthan transmission investment ranged from 18 to 28 percent; a subcomponent on reducing technical losses in distribution had an ERR of 39 to 65 percent. In addition, the transmission investment is reported to save 500 GWh a year. At the average emissions intensity of the Indian power sector, this implies a CO_2 reduction of 340,000 tons a year. Valued at $10 per ton, this would quadruple the annual stream of net benefits.

In India, owners of inefficient coal-fired power plants lack incentives to improve their equipment for two reasons. First, the regulators would require that the cost savings be passed on to consumers. Second, the utility would have to purchase power on the market while their plant is down for repairs—and the market price of power is above the artificially low *depreciated* price of old power plants. Consequently, according to the appraisal of the proposed Coal-Fired Generation Rehabilitation Project, utilities forgo opportunities to reduce coal consumption per KWh by 22 percent and realize financial returns of 28 percent. Since these returns are based on existing tariffs, the true economic rates of return are higher.

Sources: Implementation Completion Reports and Implementation Completion Report reviews; GEF 2006a.

DSM can be a cost-effective way to meet energy demand. IEA (van der Laar and Vreuls 2004) has assembled a database of DSM programs around the world. Most of these have capital and operating costs of less than €0.06 per kWh, which is significantly lower than the cost of supply in most of the utilities with such programs. These offer the potential for negative-cost emissions reductions.

However, DSM faces a fundamental incentive problem: why should a utility encourage its customers to consume *less* electricity? Utilities can be induced to do so mainly through regulation (van der Laar and Vreuls 2004). This has become more difficult to do in the wake of market liberalization and competition. Nevertheless, some states in the United States have introduced performance incentives for utilities that are able to reduce demand, and California maintains a structure where utility revenues are decoupled from electricity sales. Vermont has introduced an efficiency utility. Funded by a charge on utility bills, this nonprofit corporation is contracted by the state's Public Service Department to undertake DSM. In 2006, the cost of saved energy was 3.7 cents per kWh, against a supply cost of 10.4 cents.[3] Savings are verified and audited by the Public Service Department.

Utilities in developing countries often have an indirect motivation to promote DSM. When they are compelled to serve poor people or peak-period customers at tariffs that are below the cost of provision, the utilities can cut their losses if they can convince these customers to conserve.

Utilities have limited incentives to promote DSM . . .

except when they serve customers below cost.

Bank Engagement on DSM

The World Bank and IFC have worked on several DSM projects, in most cases with GEF funding. Table 5.2 summarizes 12 of these.

Many of the World Bank Group's DSM projects have attempted to estimate the CO_2 reductions resulting from the project. Table 5.2 provides an indication of the range of emission reductions

Table 5.2: Utility-Based DSM Projects

Project name	Country	Years	World Bank Group loan/ grant amount	CO_2 savings
Thailand Promotion of Electrical Energy Efficiency Project	Thailand	1993–98	$9.5 million GEF grant	27–45 million tons
High-Efficiency Lighting Project	Mexico	1994–97	$10 million GEF grant	763,700 tons
Poland Efficient Lighting Project	Poland	1994–98	$5 million GEF grant (IFC implemented)	3.62 million tons
Demand-Side Management Demonstration Project	Jamaica	1994–99	$3.8 million GEF grant	14,800-22,100 tons
Energy Services Delivery Project	Sri Lanka	1997–2002	$13.7 million loan plus $5.9 million GEF grant (mostly for renewable energy)	n.a.
Energy Efficiency Project	Brazil	1999–2003	$11.9 million GEF grant	1.7 million tons
Energy Efficiency Project	Croatia	2003–ongoing	$7 million GEF grant and $4.95 million loan	960,000 tons (est.)
Demand-Side Management and Energy Efficiency Project	Vietnam	2003–ongoing	$10.7 million, grant from GEF and IDA Fund	3.5 million tons (est.)
Uruguay Energy Efficiency Project	Uruguay	2004–ongoing	$6.88 million grant from GEF	n.a.
Argentina Energy-Efficiency Project	Argentina	2006–ongoing	$15.2 million GEF grant	5.9 million tons by 2012, 28.1 million by 2017, and 71.9 million by 2022
Power Sector Development Operation	Uganda	2007–ongoing	$300 million loan ($16 million of the loan is for DSM-type investments)	n.a.
Urgent Electricity Rehabilitation	Rwanda	2007–ongoing	$4.5 million GEF grant, mostly for renewable energy; the DSM component relates to studies only	n.a.

Note: n.a. = Not available.

a. From Implementation Completion Report reviews.

b. In Thailand, the utility, EGAT, funded DSM through a special, government-authorized tariff charge during the project period. Since the project ended, EGAT began funding DSM from its regular tariff revenue and funding has decreased for the most part. Thus, the regulations support DSM but fall short of requiring it.

c. In Vietnam, several laws and decrees support DSM and require the government to consider it. There appear to be no requirements for the utility to invest in DSM.

d. Uruguay is evaluating several options for regulatory support of DSM as part of the project, including a system benefit charge or an obligation to include financially attractive DSM measures in utility investment plans.

reported from these efforts. These estimates must be taken with extreme caution, however, as methodologies differ and are poorly documented, and reports of *actual* savings include projections. The estimates will be very sensitive to baseline assumptions—what kind of power source would have been used at the margin, had the efficiency program not been

Utility as DSM manager?	Regulatory requirement for DSM?	Other policy components in funding	Outcome/sustainability/ institutional development impact ratings[a]
Yes	Yes[b]	Appliance labeling, building certification, public education and awareness	Highly satisfactory/likely/substantial
Yes	No	No	Marginally unsatisfactory/uncertain/ modest
No	No	No	n.a.
Yes	No	Appliance energy-efficiency testing and labeling; capacity building	Moderately satisfactory/unlikely/ substantial
Yes	No	Design Code of Practice for Energy Efficiency, Commercial Buildings (mandatory for new buildings)	Satisfactory/likely/high
Yes	Yes	Appliance testing, certification, and labeling	Moderately satisfactory/n.a./n.a.
Yes	No	No	n.a.
Yes	Yes[c]	No, though one of the DSM components (energy efficiency in commercial facilities) implemented through a government program	n.a.
Yes	Yes[d]	Development of regulations to support DSM; assistance with incorporating energy efficiency in the overall energy strategy of Uruguay; appliance testing, labeling, and standards	n.a.
Yes (3+ utilities involved)	Under active development	Preparation of energy sector, tax, and financial policies and regulations for the promotion of energy-efficiency activities; standardization, testing, certification, and labeling program	n.a.
No	No, though the program is implemented by the government, not the utility	$80 million policy support program, including implementation of an energy-efficiency strategy and plan, as well as tariff increases (drafting the Energy Efficiency Strategy and Plan was a loan approval condition)	n.a.
Yes	No	Support of energy policy development related to DSM	n.a.

Bank-supported DSM projects use utilities as managers. implemented. For instance, the Brazilian energy efficiency program claims to have reduced CO_2 emissions by about 900 gCO_2/kWh saved; this is more than twice the emissions factor claimed by current large-scale Clean Development Mechanism projects in Brazil.

The projects have been successful in reducing energy demand and CO_2 emissions during the project term, although policy engagement has sometimes been missing or unsuccessful. The Thai project was particularly successful in transforming markets for energy-consuming products such as lighting. The current evaluation focuses on policy engagement, so it will not repeat a detailed analysis of overall project results, but rather will examine how the Bank engaged in policy discussions and how the existing policies affected the project outcomes, based mostly on documentary evidence.

Through the course of these DSM projects, the World Bank and GEF have learned about the importance of regulatory and policy support for DSM projects. (See box 5.2.) However, most of the projects have fairly limited policy and regulatory elements. And 10 of the 12 projects listed in table 5.2 have a utility as the DSM manager, even though problems can occur if utilities lack appropriate incentives, as noted in the GEF Evaluation Office's *Climate Change Program Study* (GEF 2004). Such incentive problems have restricted the sustainability and durability of DSM efforts. Most of the completion reviews and other project reviews note some problems with DSM program reductions after the projects ended.

The Project Document used to design the Mexican High-Efficiency Lighting Project (ILUMEX), for example, states that no policy or institutional reforms were needed to ensure effective project implementation. Instead of relying on utility funds or regulatory requirements for DSM, the pilot project encouraged DSM through GEF-sponsored subsidies for efficient lighting. Thus, the project was able to

---

Box 5.2: DSM in Brazil

Brazil provides a useful case study on the regulatory framework and incentives needed for utilities to undertake DSM. In 1985, Brazil established PROCEL, an agency to promote energy efficiency. In 2001, the Bank began to implement a GEF project that was designed to build capacity at PROCEL, support standards and certification development, and help to support market-oriented ESCOs. A complementary $125 million Bank loan was arranged to support 50 energy-efficiency demonstration projects to encourage demand for ESCO services.

As the project started, a severe energy crisis hit Brazil, necessitating emergency efforts in electricity rationing and efficiency, including distribution of compact fluorescent light bulbs (CFLs). These efforts succeeded in rapidly reducing demand. Yet the Bank loan was canceled because the utilities, under severe financial stress, had no incentive to promote efficiency.

Meanwhile, the GEF project focused on capacity building and equipment certification and was credited with a proportional share of PROCEL's large reported energy and CO_2 savings. But GEF support for ESCOs was cut back and the project self-evaluation noted that the project design, which assumed there would be utility demand for ESCO services, failed to take into account the lack of utility incentives for efficiency.

In 1998, after the privatization of Brazil's utilities in the mid-1990s, the regulator, ANEEL, set up a wire charge to finance energy efficiency. A 1 percent charge was added to consumer bills, and the proceeds were to be used by the utilities to promote efficiency. Instead, the utilities have used these funds for supply-side efficiency (for which they already had an incentive) or to support efficiency in public lighting (where official tariffs were low, and municipal governments often slow to pay). The utilities simply have no incentive to reduce profitable electricity sales.

Another issue both for PROCEL and for the wire charge is lack of thorough and independent monitoring and evaluation (Januzzi 2005). Although there are well-developed international standards for measuring energy savings, they were not applied in the Brazilian programs. This situation is not unique to Brazil.

Sources: Januzzi 2005; World Bank 2007c, 2007d.

demonstrate that energy savings are achievable, but because of the design and lack of built-in measures for replication, the DSM efforts essentially ended after the project was over.

A number of projects have followed ILUMEX in promoting the adoption of compact fluorescent light bulbs (CFLs), which consume only a fraction of the power of equally bright incandescent lamps. A classic example of the energy-efficiency conundrum, they typically offer high implicit rates of return, and yet are not adopted by users. A rough calculation based on current prices suggests that CFLs can save electricity at the rate of $0.01 per kWh,[4] with additional savings from the reduced need for generating capacity to serve peak demand.

One reason that consumers do not adopt CFLs or other efficiency measures is that they do not face the marginal cost of providing peak-hour electricity. This is especially perverse in the case of large commercial buildings with inefficient insulation and air conditioning systems, and is an argument for peak-hour pricing. In addition, there are information problems leading to market failure. Consumers do not trust that the light bulbs will work as advertised. They may fear, with reason, that the unstable voltage typical of many overstretched power systems will cause the relatively expensive bulb to burn out early. So one line of projects, including the IFC's Electric Lighting Initiative, seeks to certify and label good-quality bulbs or to provide warranties for their replacement. An early evaluation of the Electric Lighting Initiative estimated that the $25 million investment had catalyzed a reduction in CO_2 emissions by about 2 million tons and electricity consumption by 2.6 TWh. However, these estimates are based on assumptions about some crucial but unmeasured parameters.

The World Bank has recently sponsored or planned mass distribution of CFLs in Ethiopia, Rwanda, Uganda, and Vietnam, often as an emergency measure to address shortages of power supply. The Uganda project has demonstrated the feasibility of rapid distribution of more than half a million CFLs. Rough calculations

suggest that this reduced peak demand by 30 MW, at a cost far below that of 30 MW of additional generation (DCI 2008). The Ethiopia project contains a $1.25 million technical assistance component to assist the Ministry of Energy to study DSM. A recently approved $15 million grant project will sponsor mass CFL distribution in Argentina. This is attractive to the utilities, which are required to sell power below cost. However, the impact would be far greater if the distribution were used to facilitate an increase in Argentina's unsustainably low tariffs.

The Bank Group has supported a number of projects that distributed compact fluorescent bulbs.

The Thailand Promotion of Electrical Energy Efficiency Project, launched at about the same time as ILUMEX, proved more durable. This project also involved more extensive regulatory and policy discussions and components. The Thai utility, EGAT, found DSM to be very worthwhile because of its ability to improve EGAT's public image. There was strong government support for the DSM program in EGAT. The DSM Office in EGAT was able to successfully influence government policy—for example, the Ministry of Energy adopted new Minimum Energy Performance Standards. Thus, in many ways, this project is a good indication of how stronger engagement on energy-efficiency policy can enhance project outcomes and transform markets for energy appliances.

During the project period, EGAT funded DSM through a tariff surcharge; the regulator supported but did not require this. After the project ended, EGAT eliminated this surcharge and began funding DSM through its regular tariff revenue, but DSM was rarely funded at the allocated level. EGAT still maintains its DSM program, 10 years after the project closed, but the program's future is not entirely certain. EGAT is undergoing restructuring and privatization, and the DSM Office no longer has a strong advocate in EGAT.

Overall, the Thai DSM program has been very successful, in part because it has been able to involve key stakeholders in ways that highlighted their own self-interest in participating. However, such stakeholder involvement can take time, and

Many DSM programs have no regulatory requirements for demand management, and monitoring and evaluation are weak.

without external funding, the benefits of initial stakeholder involvement can fade if there is no clear mechanism to ensure funding for DSM.

The DSM project in Vietnam has many elements similar to those of the Thai DSM project. The government of Vietnam drew from Thai examples in drafting its legislation and decrees to support energy efficiency. Because this project was launched in 2003, it is too early to say if it will be sustainable in the long term, but to date, progress seems impressive. As in Thailand, there is no firm requirement for the utility to invest in DSM. Now the utility is very much in favor of DSM because it is reducing the utility's losses for electricity sold to customers eligible for below-cost electricity rates. The project does not explicitly fund policy-related work to support DSM.

Many other DSM programs with World Bank support have no regulatory requirements for DSM. This includes several of the most recent DSM-style projects. Utility DSM funding has at times been reduced during or after a World Bank project when the utilities that fund DSM find other priorities. An example of this is the Jamaican Demand-Side Management Demonstration Project, where the local utility used a large portion of its own funds, initially designated for DSM, on emergency power plant repair.

Monitoring and evaluation are generally weak, but there are signs of improvement. The CFL projects are of special interest in this regard because they are potentially highly amenable to monitoring and because evaluation could answer a number of critical questions for policy and program design. These include the degree to which free or low-price distribution induces consumer willingness for subsequent commercial purchase and the degree to which consumers take advantage of efficient bulbs through increased lighting hours rather than reduced electricity consumption. The Electric Lighting Initiative evaluation stressed the need to include better planning for rigorous data collection from

the start. This has not proved possible in emergency-driven projects. However, the current CFL project in Ethiopia incorporates a randomized control trial impact analysis. And with the advent of programmatic CDM (under which these projects could be presented for carbon finance), much more rigorous monitoring could be brought to bear. In India (unconnected to the Bank), an innovative monitoring effort sponsored by the Bureau of Energy Efficiency will use automatic wireless data reporting from a sample of residences to measure the impact of CFLs on electricity consumption.

Lessons Learned

Most of the Bank's DSM efforts are investment-focused. Thus, they achieved meaningful energy efficiency gains during the project period, but the projects did not typically result in new legislation or regulation that would provide ongoing financing for DSM. Some CFL distribution projects have included standards or certification components, whose long term-effect is yet to be seen.

The Bank has consistently partnered with utilities—rather than regulators or energy ministries—in supporting DSM programs, often building on existing relationships. Utilities may have the capacity to implement these programs, but their incentives to do so are limited to specific market segments or situations. The Bank Group's current emphasis on energy finance and ESCOs can be seen as a way of promoting DSM while bypassing engagement with utilities or regulators. However, global experience suggests that regulatory drivers of DSM can complement ESCO market expansion.

Successful DSM programs benefit from well-designed systems for monitoring and verification of energy savings. While such requirements can be adopted at the utility level, policies and regulations can also provide support for robust monitoring and verification systems.

Appliance Standards and Building Codes

Appliance standards and building energy codes have proven to be some of the most effective and cost-effective policies for improving energy efficiency globally. A review of U.S. experience with appliance standards (Gillingham, Newell, and Palmer 2006) found estimates of the net benefits of appliance labeling of $56 to $196 billion over 25- to 30-year periods. A self-evaluation of the Thai energy efficiency project, which emphasized labeling of high-efficiency lights, refrigerators, and air conditioners, found savings of 28 TWh over 1993–2004 and projected savings of 61 TWh over 2004–10, arising from a $40 million project (GEF 2006).

Building energy codes are important because of the long-term and significant impact they can have on reducing energy demand. Buildings typically last for 30 to 40 years. The initial design of buildings is the single most important factor in determining their energy consumption pattern. Energy savings measures are less expensive during the initial construction than through retrofits later on. But builders rarely have an incentive to maximize efficiency because they do not pay the energy bills, and buyers have little way of knowing what future energy performance may be. Globally, buildings are responsible for 15.3 percent of GHG emissions—more than the transport sector (Baumert, Herzog, and Pershing 2005). The construction boom in fast-growing economies such as China and India presents an opportunity to adopt high-efficiency energy standards in order to change demand trajectories for the life of these new buildings.

Appliance Standards and Labels

The term **appliance standard** is often used to describe two different, though related, types of policies. The first sets minimum efficiency levels for appliances such as refrigerators, lighting ballasts, boilers, hot water heaters, and air conditioners. The second type of policy involves appliance labels (voluntary or mandatory) that describe energy performance and consumption or endorse a product as energy efficient. Testing laboratories and protocols are essential in implementing appliance standards and labeling programs.

Building Energy Codes

There are two main types of building energy codes: prescriptive and performance-based. Prescriptive codes specify the characteristics of building components—regulating, for instance, the insulation efficiency of windows or walls. Performance-based codes set an energy budget for a whole building. This allows building designers the flexibility to trade off different types of components. Enforcing prescriptive codes is typically easier than enforcing performance-based codes, because inspectors can check the specific components against a set standard. Performance-based codes offer potentially greater cost-effectiveness, but require the development and use of software that can model the building's total energy use.

Because enforcement of building energy codes is necessarily local, and compliance requires checks at the building level, strong capacity and adequate staff are needed. Inspections and compliance checking may be done either by the entity that checks for compliance with other types of building codes (such as codes for structural integrity and fire safety), or in some countries, by specialized building energy units. Each approach has advantages and disadvantages. Testing laboratories and procedures are also essential to independently determine the performance of building materials and components such as windows and insulation.

Many developing countries have building energy codes, but compliance systems may be weak. Where staffing is inadequate, the inspection system may rely on checking plans rather than implementation.

Bank Engagement on Standards and Codes

The World Bank has engaged on codes and standards several times during the past 15 years. This work is not as broad or robust as the work on DSM or tariff policy, so there is not as much evidence to examine. Most of this work has been done as relatively small components of other DSM or energy-efficiency projects. Table 5.3

Appliance standards and building energy codes are among the most effective and cost-effective policies for improving energy efficiency globally.

Table 5.3: Projects with Appliance Standard and Building Energy Code Components

Project name	Approval year	Component funding (total funding)	When code adopted	Description of component
Thailand Promotion of Electrical Efficiency Project[a]	1993	> $1.67M ($9.5M)	During project (later adopted by government too)	• Appliance labeling (by utility) • Development and promulgation of building and appliance codes in order to enforce minimum efficiency standards (done by utility) • Establishment of testing laboratories
Jamaica Demand-Side Management Demonstration Project[a]	1994	$0.6M ($3.8M)	Before project; strengthened during project	• Strengthening of capability of Jamaica Bureau of Standards • Enhancement of testing laboratory capabilities • Campaign to promote voluntary building code and appliance labeling program
Brazil Energy Efficiency Project[a]	1999	$3.4M ($11.9M)	Before, though enhanced law covering standards adopted during project	• Support of utility-funded standard and labeling program • Definition of energy-efficiency standards to comply with efficiency law (this component later canceled because of poor consultant performance) • Strengthening of capacity of testing laboratory
Uruguay Energy Efficiency Project[a]	2004	> $1M ($6.88M) Funding for standard component decreased after project start	n.a.	• Appliance testing program • Labeling and standards program including a voluntary energy-efficiency seal for main household appliances, lighting equipment, building thermal envelope, and industrial and other equipment and materials
Argentina Energy Efficiency Project[a]	2008	> $3.7M ($15.2M)	During project	• Comprehensive program for energy-efficiency standards and labeling of key equipment, including appliances, industrial equipment, and building materials • Modernization of certification laboratories • Strengthening of capacity of standardization bureaus • Regulatory and enforcement activities
Sri Lanka Energy Services Delivery Project[a]	1997	> $1.9M ($5.9M grant + $13.7M loan)	During project	• Design Code of Practice for Energy Efficient Commercial Buildings (mandatory code, written and adopted during project) • Development of institutional capacity in the energy-related public and private sectors to incorporate the Code of Practice into building design and operations and to monitor the energy saving
China Heat Reform and Building Energy Efficiency Project	2005	> 0.8M ($18M)	Before project, strengthened during project	• Technical studies on developing more stringent code • Development of code compliance enforcement capacity

Source: Project documents.

Note: n.a. = Not available. Unless noted otherwise, all funds are from GEF grants. Numbers shown as less than the listed amount mean that project documents combine several project components into one budget line, making it impossible to determine how much is spent on codes and standards alone. In most such cases, it appears that the spending on the codes or standards component is less than half of the figure given.

a. Project was also reviewed in the DSM chapter.

highlights projects with appliance standard or building energy code components. No projects focus exclusively on these issues, and even among the projects in the table, there are several where the code and standard component was not actually funded directly by the Bank or GEF, but rather with local resources.

Overall, the World Bank and GEF work on appliance standards and building codes is very successful in adopting new codes and standards, and somewhat less successful in establishing the necessary infrastructure to implement the codes. Funding is the key limiting factor, both with assistance on code adoptions and with capacity building for enforcement. In evaluating the Bank's experience in more detail, it is helpful to look at it from three perspectives:

1. Results in assisting countries in adopting new codes and standards
2. Experience in helping countries develop stronger enforcement systems to make the codes work in practice
3. Monitoring and learning from experience.

Table 5.3 indicates that the Bank has had much success in its work to assist countries in adopting new codes and standards. In almost every case, the countries have adopted new regulations or strengthened existing code and standard systems. This is an extremely high success rate for engagement on any policy, and it is particularly noteworthy given the Bank's low level of funding for code and standard development. By 2004, 74 countries had adopted codes or standards of some kind.[5] The challenge for these countries, and for the Bank, is to create the institutions that will oversee the effective implementation of these standards.

However, regardless of how large or small a country's codes and standards programs are, the largest costs associated with the programs are for implementation.

Codes and standards components of Bank projects are often envisioned as a tool that can enhance the ability of utilities to implement DSM

programs (all but one of the projects described in table 5.3 have a major DSM component). This can be useful and effective, in that the DSM organization can help ensure that there is a wide market—for compliant appliances, for example. The Thai project provides an excellent example of this. However, this approach can also be limiting if the codes and standards are written by or for the DSM program instead of for the country as a whole. When codes and standards are written for a DSM program, compliance systems outside the program are often weak or nonexistent. So while this situation is better than having no code at all and it may eventually lead to a broader national compliance system, it may also create vulnerabilities by linking so much to a single DSM entity.

The Bank's work on codes and standards is not as broad or robust as that on DSM or tariff policy.

Bank and GEF efforts have generally succeeded in the adoption of new appliance standards and building codes, but have been less successful in establishing the infrastructure needed to implement them.

Aside from voluntary efforts to use standards and labeling as DSM tools, the Bank's main engagement on code and standard implementation has been in partially funding testing laboratories. All of the projects with appliance standard components have worked to develop testing laboratory capabilities. Except for the China Heat Reform and Building Efficiency Project, there is little evidence of project-based work to enhance other types of inspection and enforcement systems or to build government capacity for disseminating the code. In other words, the Bank's efforts have done relatively little to create capacity for enforcement of mandatory codes and standards. Interestingly, one of the concerns that Bank staff have expressed about work on mandatory codes and standards in general is that codes and standards are difficult to enforce. Yet one could argue that this is a self-fulfilling prophesy if enforcement systems are not included in the program design from the beginning. One of the problems in this regard is the level of funding. At current levels, it is not enough to engage on both the development of regulations and the capacity building to enforce them.

Monitoring and learning from experience is also very important, particularly in a developing area such as codes and standards. While programs in the West have been extensively evaluated and modified based on these evaluations, there has been much less monitoring and evaluation of codes and standards programs in developing countries. This is in part because these programs are younger, but also because the programs are stretched to deal with enforcement, and monitoring may drop in the list of priorities of a poor country.

There is very little World Bank documentary evidence on the success or failure of the codes and standards components of Bank projects. Monitoring and evaluation of this kind of effort cannot end with project closure, but requires tracking of standard adoption and implementation. Aside from the Thai project, there is very little evaluation of the energy and emission results of the codes and standards. The Thai labels, for example, have reduced annual electricity consumption by 1,200 GWh. For most projects, there is no information on the estimated CO_2 emission reductions from the standards, labeling, and/or building energy code components. Calculating these emission reductions is relatively easy once the energy savings are determined, given information about the source of electric power.

There is little documentary evidence regarding the success of the codes and standards components in Bank projects.

The lack of documentary evidence is most likely linked to financing: the codes and standards work received only a fraction of the financing in any given project, and was thus not the priority of assessments at project close. The lack of evidence on the results of these project components also makes it difficult to learn from them.

One point that does come through from the documentary evidence is that funding for the codes and standards components was reduced in several cases. This is true in both Uruguay and, to a certain extent, Brazil. There is no reason given for the funding reduction in the Uruguay project. In Brazil, overall, the testing, certification, and labeling component of the project was given greater emphasis when the project was restructured. However, a subcomponent most closely linked to code and standard development was canceled: the problem was a poor-quality report prepared by a consultant.

In several cases, funding for the components was reduced.

In the case of the Jamaica DSM project, the Jamaica Bureau of Standards had initially requested a higher level of funding for testing and building capability to handle the country's new appliance standard program, but ultimately this was not considered a priority for DSM. The Project Appraisal Document also mentions a concern that the Jamaica Bureau of Standards might not be able to test equipment for the DSM program fast enough; the project contingency plan for handling this risk was to test equipment in the United States. The Jamaica DSM project did build some lasting capacity and testing capabilities at the Jamaica Bureau of Standards, but the country was clearly willing to go farther during the project.

The designers of the Vietnam DSM project actually considered including a component on codes and standards, but this was not part of the final project design. The Project Appraisal Document notes:

The project considered additional efforts to support the codes and standards work initiated under the SIDA-supported first phase. However, given the very low demand for energy efficiency equipment at present, combined with the limited government capacity to test and enforce national standards, it was determined that an initial focus on creating greater market demand for energy efficiency equipment would be a more appropriate priority at this stage. As the program and markets develop further, the appropriateness for national standards and codes would improve as well as the prospects for successful introduction and implementation.

This excerpt reflects the view that codes and standards are unlikely to transform markets. Experience from around the world indicates that this is not the case.

Clearly, the engagement on codes and standards to date has been very small, and such small projects can be difficult to implement at the World Bank. The question, then, is: are there ways to structure projects that involve building energy codes and appliance standards that might be better suited to the World Bank's structure? If implementation is a greater focus, projects or project components related to codes and standards will necessarily become larger.

The Heat Reform and Building Energy Efficiency Project in China (see box 5.3) can provide insights into how projects might address implementation needs. This project involves working with local authorities to design better building inspection procedures and capacity; it also involves a significant investment component related to improving energy efficiency in existing buildings. Working with local builders and helping them to improve new buildings to meet the code may also be a useful approach to enhancing understanding and enforcement of the code, and for ensuring that the code takes builders into account.

Public Buildings

The World Bank is also considering expanding its work on energy efficiency in public buildings. To date, the Bank has engaged in two or three such projects. The largest (in funding) was the Kiev Public Building Energy Efficiency Project, approved in 2000 with an $18 million World Bank loan. This project was rated as satisfactory. The Bank is also investing $12 million in energy efficiency upgrades in state hospitals and schools under the Serbia Energy Efficiency Project, approved in 2004. In addition, Argentina has a nascent government program to promote energy efficiency in public buildings. While the GEF's Energy Efficiency Project devotes some technical assistance to that program, the project's efficiency fund targets small and medium-size enterprises. This focus was questioned in a project design (or STAP) review, which suggested:

Devoting some funds to developing the demand for ESCO services in a few key sectors such as in large office buildings and in the public sector. These sectors are typical

---

Box 5.3: Heat Reform and Building Efficiency in China

Heating efficiency in China's colder northern areas is a matter of national economic concern with global implications. Housing is expanding rapidly. It is expected that 6 billion square meters of new space will be erected over the next 20 years. Heating these buildings consumes an inordinately large quantity of energy, most of it from carbon-intensive coal consumption.

There are interlocking reasons for this inefficiency. On the demand side, incentives are askew. Heating costs are paid by employers, so households have no incentive to control heat. Heat is billed at a flat rate, so no one has an incentive to reduce heat at the margin. Even if they wanted to do so, heat is generally not controllable at the household level. On the supply side, materials and techniques fall short of what is technically and economically feasible, and existing building codes are imperfectly enforced. Progress requires advances on all fronts, since there will be little demand for better insulation without price incentives, and little appetite for assuming price responsibility without improved heating efficiency.

The Bank has had a long interaction on these issues with Chinese authorities, dating back at least to 1990. Energy efficiency, including residential heat efficiency, was stressed in the 1994 study *China: Issues and Options in Greenhouse Gas Emissions Control,* undertaken jointly by the Bank, the United Nations Development Programme, and the Chinese government (National Environmental Protection Agency of China and others 1994). The World Bank has been involved in projects for the promotion of energy-efficiency finance and the introduction and diffusion of efficient boilers. Dialogue on efficiency issues continued, and two studies on building efficiency (in 2000 and 2002) provided inputs for policy setting. Trust fund support, including that of ESMAP and the Asia Sustainable Alternative Energy Program (ASTAE), was crucial to maintaining a stream of formal and informal studies and dialogue. With these inputs, Tianjin emerged as a city interested in innovating in heat policy and building standards. The GEF/World Bank China Heat Reform and Building Efficiency Project, initiated in 2005, supports Tianjin as a demonstration center for these reforms, with components to support replication in other cities and to support capacity for national policy making in this area.

markets for ESCO services in other countries. Some funds could be used to promote use of ESCOs in these sectors, publicize the results of demonstration projects, and if necessary reform government procurement rules to enable performance contracting and use of ESCOs by the federal, state, and local governments. The public sector often lacks the capital to make energy-efficiency investments on its own, and thus is an excellent market for ESCOs if third-party financing is available.

The response to this review defended the project's main focus on small and medium-size enterprises as a lower-risk area for ESCOs. But from a policy perspective, what is important is finding where market failures are greatest and addressing them in a sustainable way.

The potential for energy savings (and thus lower GHG emissions) in the public sector is great in most countries. Governments have more control over their own energy use than they do over energy use in the broader economy, and governments are often among the largest energy consumers in a country, given the scope of their activities.

The potential for energy savings in the public sector is great in most countries, but there are significant challenges to realizing that potential.

Still, the public sector presents unique challenges. For example, public entities may not have the power to reallocate their budget to energy-efficiency investments, so financing is essential. ESCOs can often play a positive role in this area. Public entities may not be allowed to use future energy savings: their budgets may be reduced to cover only actual energy costs, which reduces incentives (and creates challenges for repayment). Procurement rules may force government agencies to award contracts to the lowest bidder, without considering life-cycle costs.

Many developing countries have begun to address these issues. For example, China has developed an energy-efficiency procurement program, with a list of qualifying energy-efficient products that receive preferential treatment in procurements. Russia and Ukraine have both adopted programs to help finance energy-efficiency improvements in state-owned facilities, based largely on the U.S. Federal Energy Management Program. India is trying to promote energy efficiency in new government buildings by ensuring that these buildings meet or exceed the new voluntary energy code for commercial buildings. The plan is to use this effort to spearhead nationwide implementation of the new code (APP 2007; PNNL/ARENA-ECO 2003). This growing interest among developing countries creates an excellent opportunity for the Bank to engage constructively in this area.

District Heating

District heating has also been an important area of engagement for the World Bank, with total commitments of $1.8 billion. Much of this investment has gone to supply-side efficiency: the replacement of inefficient, polluting boilers.

The Bank began working on district heating in the early 1990s, after the fall of the Berlin Wall created new opportunities for engagement in the former Eastern Bloc. District heating is a very important form of energy in Eastern Europe and the former Soviet Union. It provides up to 70 percent of residential space heating, with the more northern countries typically seeing the largest shares of heat from district heating. The Soviet-designed systems were inefficient compared with the district heating systems in the West. They did not have adequate controls and were often oversized. They also relied less on combined heat and power production, which is typically very efficient, than was rational given the concentration of heating demand that the systems created.

The Bank has undertaken 41 district heating projects since 1991. Some of the projects involved policy elements, either at the local or national level. For example, the Bank encouraged tariff increases and reform and restructuring of systems to make them more commercially oriented. In some cases, as in Poland and Romania, policy engagement encouraged governments to take a broad look at integrating district heating in the overall energy policy and strategy. These were all positive steps.

However, there were also cases where the Bank's policy advice and focus may have been too narrow, which created problems in the long term. One of the most important examples of this related to demand. Demand for district heating sometimes dropped dramatically following the introduction of reforms. To some extent, this shift was a natural decline linked to structural shifts (industrial demand dropped particularly fast). However, the extent of decline went beyond this in many countries. There were many factors behind this, including the poor quality of the service during interim years and rising heat tariffs, which encouraged efficiency. Rising district heating prices relative to the prices of other heat alternatives, such as natural gas, were also a major issue. This created a market imbalance and encouraged customers to disconnect from district heating. This illustrates the importance of coordinating price reforms across competing fuels.

Clearly, district heating in transition economies went through major changes from 1989 to the present, most for the better. Demand has begun to increase again in most countries and customers are starting to return. At the same time, an overly narrow focus on reducing subsidies and improving supply may have missed opportunities to help systems adjust during the transition with less destructive declines in demand.

Today, many of the customers who switched away from district heating are finding that their natural gas bills are very high. The poorest customers have struggled to find alternatives when district heating systems have collapsed—for example, in Romania.

In its more recent engagements, the Bank has taken a broader approach—for example, ensuring that natural gas reforms take district heating into account and vice versa. A recent project in Kazan, Russia, included comprehensive collaboration with the city in improving communal and housing services; the Bank collaborated closely with the city on fiscal, administrative, and pricing reforms. As a result,

the city has reduced its heat subsidies, created a targeted poverty benefit for the poor, improved the fiscal position of both the city and the district heating company, and improved district heating services (IEG 2008c; World Bank 2008d).

In some cases the Bank's policy advice and focus may have been too narrow.

The other place where the Bank has invested significantly in district heating is China. China has the second-largest district heating sector in the world, and, unlike in Russia, where demand is growing moderately now, demand is growing quite quickly in China. In response to Chinese interest, the Bank has had a deep and broad district heating policy dialogue with China for most of the past decade. Major policy reforms have grown out of this dialogue. The reforms consider the need to raise tariffs to cost-recovery levels; to ensure that consumers are responsible for their own bills (and not their employers); and to provide better controls and metering, paired with consumption-based billing (instead of billing based on apartment size).

A large share of district heating investment is linked to consolidating small, inefficient (and polluting) municipal boilers. Such consolidation makes sense, but from an efficiency perspective, it would make even more sense if the new supply came from combined heat and power plants and not heat-only boilers. Only one of the Bank's six district heating investments in China has involved new combined heat and power plants. Such production is more efficient than separate power and heat production, and it reduces emissions by half because the heat (in the form of steam) is first used to generate power, and then the exhaust heat is used as heat supply for district heating or industry, rather than being wasted (as in a single-purpose plant).

Conclusion

Energy-efficiency efforts—at the Bank, and arguably in the world at large—consistently fall short of the level suggested by rhetoric and analysis. At the Bank, a small group of dedicated enthusiasts has pursued energy-efficiency projects, despite an incentive structure that does not favor small, staff-intensive projects that require sustained, long-term engagement with clients. In this they have been supported by trust fund sources such as the Asia Sustainable and Alternative Energy Program and ESMAP. Projects have relied heavily on GEF support, suggesting that concessional resources were important in securing client interest. At the same time, few energy projects have had strong policy components. Among the projects with policy components, many involved partnerships with utilities that had sharply constrained interest in promoting efficiency.

Discussions with staff and other stakeholders are consistent, with some standard diagnoses about the neglect of energy-efficiency opportunities. Energy efficiency is simply not as visible as energy generation. It is difficult to spend large sums of money on energy efficiency quickly (except with the mass distribution of CFLs or in some supply-side projects), and yet energy-efficiency projects are often complex or difficult. This makes them less attractive to managers and

Internal and external incentives favor supply over efficiency.

agencies that use disbursements as a measure of action and large turbines as a visible symbol of achievement. Energy efficiency is viewed by some as being less real than generation, although numerous analyses show that much of the demand for energy services over the next 30 years can be provided more cheaply through increased efficiency than through increased generation. A lack of rigorous monitoring and evaluation reinforces skepticism about the true magnitude or cost of achieving efficiency gains. IEA (2008b) notes the large gaps in energy efficiency indicators that countries could use to diagnose areas of opportunity and to track progress.

Yet it is worth stressing that there is client willingness to engage on the issue of efficiency policy. As noted, the country strategies for many of the Bank's clients with large or inefficient energy sectors include efficiency objectives. Many countries have adopted national energy efficiency policies. Prominent examples include India's Energy Efficiency Act (2001) and China's goal of reducing energy/GDP by 20 percent between 2005 and 2010.

Chapter 6

Evaluation Highlights

- Gas generation is more flexible, has lower environmental cost, and is easy to install, so the main barrier to the use of natural gas is its availability.

- The flaring of gas associated with oil production wastes energy and releases large amounts of CO_2 into the atmosphere.

- In many cases it makes economic sense to recover the associated gas.

- The Global Gas Flaring Reduction Partnership has had some success in promoting dialogue, raising awareness, and developing and disseminating knowledge, but flaring remains at high levels.

- The ERR to the use of associated gas is high, but financial rates of return are strongly affected by pricing policies.

Gas flaring and pipeline equipment, SASOL Pipeline, Sub-Saharan Africa. Photo courtesy of SASOL/IFC.

Natural Gas Flaring

When it comes to generating power, natural gas is more appealing than its main competitor, coal. Gas burns cleaner, without spewing lung-damaging particulates and contributing to acid rain.

Gas plants are much cheaper and faster to construct than coal-burning plants—important considerations for private investors—and can be used for baseload or peak power. And, of crucial importance to the topic at hand, a modern gas combined-cycle turbine power plant emits only about half as much CO_2 per kilowatt-hour as a coal plant.

Promotion of natural gas for power would seem to be an attractive win-win policy for climate mitigation. The technology is proven, and the potential scale is large. What are the barriers? In brief: geography, which has scattered gas deposits quite unevenly across the planet; lack of infrastructure to transport the gas from its often remote origins; and policy that shapes the incentives for extraction, transmission, and use.

We focus here on policy, with particular attention to the problem of gas flaring. *Associated* gas, a by-product of oil production, is often vented or flared (burned at the wellhead) instead of being captured and transmitted. The scale of flaring is immense: about 160 cubic kilometers per year, containing enough energy to power Sub-Saharan Africa twice over. The annual flux to the atmosphere could be more than 400 million tons of CO_2e—about 1 percent of the global total. The logistical and incentive problems of capturing and tapping this energy are illustrative of wider policy issues.

Context

Utilities continue to opt for gas-powered plants where gas is available, even though some calculations show coal plants to have lower average generation costs. Figure 6.1 shows a breakdown of recent, current, and planned power plant capacity for two groups of countries: those with gas access but no coal reserves, and those with access to both. Countries with gas but no coal continue to opt for gas and hydropower, even though coal is transportable. More surprising, even countries with coal reserves are putting 29 percent of new capacity into gas versus 21 percent for coal. There are large disparities within this group: China, Indonesia, and India continue to rely on coal, while Russia and Kazakhstan emphasize gas. The numerous, but mostly small, countries with neither coal nor gas opt mostly for oil-fired power plants. (Only seven countries had coal reserves but no gas access.)

Where gas is available—even in some countries with coal reserves—it is often the fuel of choice to generate electricity.

Figure 6.1: Recent and Planned Generation Capacity Additions by Fuel Type

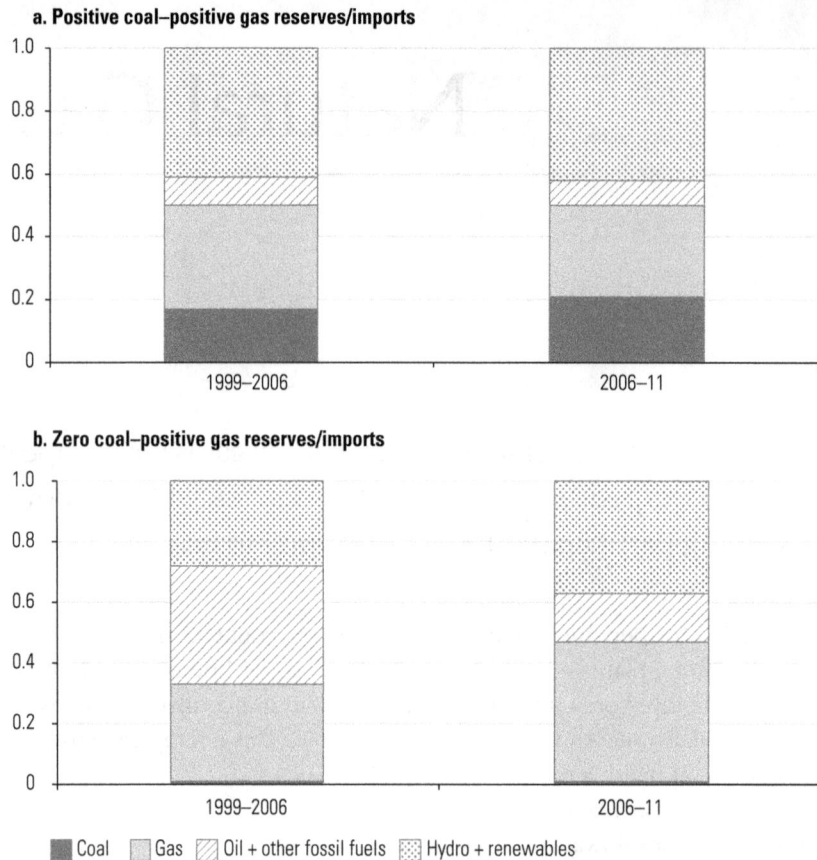

a. Positive coal–positive gas reserves/imports

b. Zero coal–positive gas reserves/imports

Coal ▪ Gas ▪ Oil + other fossil fuels ▨ Hydro + renewables ▨

Source: Meisner (2008) based on Platts World Electric Power Plant database.

Costs of gas generation are slightly higher than costs of coal generation. A detailed study by the National Energy Technology Laboratory (NETL 2007) shows capital costs for gas to be about 35 percent of those for a comparable subcritical or supercritical coal plant, with total levelized electricity cost about 6 percent higher. ESMAP (2007), however, reports a 24 percent differential in levelized cost. These 2007 calculations are already out of date because of changes in energy and capital costs, but coal prices are rising more quickly than gas prices at this writing. And Blyth and others (2007) reckon that gas is preferred even when the price per energy unit is twice that of coal.

And compared with coal, gas also offers local and

global environmental benefits. The National Energy Technology Laboratory study estimated that a 600 MW coal plant would emit 211 tons of particulates and 1,400 tons of SO_2 annually, with emissions controls in place. However, many plants in developing countries do not have such controls. In contrast, particulate and SO_2 emissions from gas plants are negligible, and NO_x emissions are only about 10 percent of those from coal.

The CO_2 differential is very large. For the 600 MW plant, emissions would be 1.5 million tons from gas, but 3.3 to 3.5 million from coal. The differential remains even when life-cycle emissions are factored in for LNG (liquefied natural gas), which requires energy-consuming liquefaction. Hondo

(2005) considers coal transport, LNG liquefaction, transportation, and leakage, and finds that life-cycle emissions per kilowatt-hour are still 47 percent lower for gas than for coal.

Given the superior flexibility of gas, lower local environmental costs, and ease of installation, a major barrier to its use is physical availability. Gas deposits are highly concentrated in a few countries, and transport requires expensive pipelines or liquefaction facilities. From an energy security standpoint, consuming nations are concerned about reliability of supply in a thin market. So increasing the supply of gas in areas that would otherwise depend on coal or oil is a win-win approach to increasing electricity availability while reducing GHGs.

Gas supply depends on exploration and field development, infrastructure construction, and policies that regulate and motivate gas development, transportation, and use. While the Bank has had a role in pipeline construction, here we focus on its involvement at the policy level, particularly with regard to gas pricing and regulation.

The Paradox of Gas Flaring

Oil wells sometimes spout dissolved gas. Some of this associated gas is captured and used productively—burned for power or reinjected into the earth to prime more oil production. But over the period 1995–2006, an estimated 160 billion cubic meters (bcm) per year were flared (Elvidge and others 2007). If used for power generation, this gas could have produced about 6.75 TWh of electricity annually, nearly twice the current output of Sub-Saharan Africa. If delivered to markets at current world prices, the value of this gas would be about $60 billion per year. But instead, flaring releases the equivalent of more than 400 million tons of CO_2 into the atmosphere each year, and soot from incomplete combustion adds an additional warming load. In addition, an unmeasured amount of gas is simply vented, with 25 times the warming effect of CO_2.

Based on satellite observation (Elvidge and others 2007),[1] the largest source of this flared gas is Russia, with an estimated 51 bcm[2] in 2004, followed by Nigeria (23), Iran (11), and Iraq (8). Another 18 countries each flared more than 1 bcm (enough to power an 850 MW power plant).

The cost of gas generation is slightly higher than for coal, but it offers environmental benefits over coal.

Why waste valuable fuel? A basic question is whether it makes economic sense to collect, compress, and transport the gas. For wells that are scattered, small, and far from pipelines or electricity consumers, it will not. Nonetheless, governments may restrict or prohibit such flaring on environmental grounds. These restrictions impose costs on the oil producer and owner and require the will and capacity to enforce them on the part of the environmental authorities.

Significant amounts of gas associated with oil are simply burned off, wasting the energy and releasing large amounts of CO_2 to the atmosphere.

However, in many cases, economic fundamentals would support the recovery of associated gas—if gas were valued at world market levels or at the cost of the alternatives available to local gas users. So the persistence of gas flaring suggests a combination of technical, regulatory, market, and policy failures.

The Global Gas Flaring Reduction Partnership

The Bank-led Global Gas Flaring Reduction Partnership (GGFR) was initiated in 2001 to "support national efforts to use currently flared gas by promoting effective regulatory frameworks and tackling the constraints on gas utilization." A public-private partnership, its members include governments of 14 gas-producing countries and regions, 10 oil companies, the World Bank Group, and OPEC. Its budget was $1.5 million in 2007, rising to $3.5 million in 2008, and it is supported by a number of national donors.

In many cases, economic fundamentals suggest that recovery of the associated gas would make sense.

Some initial studies (Gerner, Svensson, and Djumena 2004) diagnosed several generic problems:

- ***Inadequate technical practices and regulation of flaring operations***—Some operators may lack

81

technical expertise in handling flares. Regulatory agencies may lack the knowledge and resources to set and enforce rules.

- **Poor contractual arrangements**—For instance, production-sharing contracts (between governments and oil producers) may not allow producers to recover the costs of collecting and transporting associated gas. Or governments may have legal rights to associated gas, but no ability to use it.
- **Inadequate pricing or access policies**—Legal or regulatory caps on gas or electricity prices may dampen incentives to recover flared gas. Subsidies for alternative fuels could have the same effect.

To address these barriers, the GGFR set up the following objectives and lines of action:[3]

- Develop and promote voluntary standards on flaring practice.
- "Survey and establish regulations followed by disseminating upstream regulatory best practice," where "regulation" refers narrowly to flaring and venting practice rather than broad sectoral policies.
  - Help to "realize gas flaring reduction projects by establishing appropriate incentives mechanisms (carbon credits for lowered emission, establishment of methodologies) leading to a reduction of financial barriers. Carbon credits will be utilized, where feasible, as a possible incentive to develop, especially, marginal fields."
- "Facilitate commercialization of otherwise flared gas in GGFR focus countries through identification of projects and reduction of barriers. This includes achieving access to international markets, local/domestic market development, and small-scale gas use, especially for remote areas and marginal developments."

The GGFR has promoted dialogue, raised awareness, and developed and disseminated knowledge.

But endorsement of and adherence to the flaring standard that it helped to develop have been below expectations.

The GGFR has promoted dialogue, raised awareness, and developed and disseminated knowledge. With a direct membership of 14 countries, the partnership now comprises territories

responsible for about 50 percent of flaring. It has sponsored stakeholder dialogue in a number of countries. It has produced informative studies on the state of gas flaring regulation, the causes of gas flaring, and methodologies for assessing the potential of flaring projects to use carbon finance.

The GGFR, working in consultation with partners, published a Voluntary Standard in 2004. At its core is a commitment to eliminate "continuous flaring and venting of associated gas, unless there are no feasible alternatives." Those endorsing the standard commit themselves to develop and implement action plans and to "regular reporting of flaring and venting levels and progress on implementation," with public reporting required within two years after adoption. GGFR regards the consensus-creation of the standard to be a significant accomplishment.

However, endorsement of and adherence to the standard have been below expectations. The standard has been officially endorsed by all the GGFR's international oil company partners, but by only four national partners: Algeria, Cameroon, Chad, and Nigeria. However, four countries (Equatorial Guinea, Kazakhstan, Nigeria, and Qatar) have deadlines for zero flaring. Only one company and one country have adopted formal implementation plans for gas flaring reductions that are consistent with the Voluntary Standard, although additional partners have similar programs in place. While all companies are reporting flaring and venting data to the GGFR, only four countries (Cameroon, Canada, Norway, and the United States) have reported flaring data for 2006, and venting has been reported only by Canada and Norway. These reports are not publicly disseminated by the GGFR.

Slow progress on reporting undermines the flaring reduction agenda and points to deep-seated issues. Reporting has been shown to be a key feature of other voluntary environmental standards, and accurate data are essential for tracking progress toward reduction goals. Measuring flaring and venting at the wellhead is technically difficult and expensive, and compilation of data across many producing locations requires standardized procedures. Recognizing this, the GGFR developed and tried to popularize a data tool for reporting. The failure of this tool to find takers suggests nontechnical barriers to reporting. Given the legal penalties and social disapproval associated with flaring, the costs of reduction (which may include reduced oil production in some cases), the expense of measurement equipment, and the weak capacity of regulatory agencies for monitoring and enforcement, some oil producers do not have strong incentives for accurate reporting of flares and vents.

Against this context, the GGFR has invested in an innovative alternative for monitoring flare volumes and locations: the use of remote sensing. A GGFR-sponsored study used nighttime satellite imagery to detect global flaring activity over 1995–2006 (Elvidge and others 2007), with updates in progress. While these estimates are themselves subject to measurement errors, they provide a useful cross-check on reported volumes, and for some areas may constitute the only available data. The greatest disparity between official reports and the satellite data is for Russia, where the satellite observations suggest a much larger volume than reported. The publication of these reports in May 2007 followed closely on then-President Putin's state of the union address, which declared flaring to be an "unacceptable waste" and cited a flaring volume higher than previous official reports.

The GGFR has worked with partners to facilitate commercialization of currently flared gas. It has encouraged multilateral discussions on commercializing flared gas in the Gulf of Guinea and has supported economic analyses of commercialization constraints and possibilities in Nigeria, Russia, and elsewhere.

The GGFR has devoted considerable effort to promoting the use of carbon markets to reduce flaring. The underlying idea is that use of associated gas may, by itself, provide only marginal returns and may therefore not attract investment by oil producers with more remunerative opportunities. But use of associated gas can also reduce GHG emissions. These reductions, worth money on carbon markets, could tip the financial balance toward gas recovery. For instance, the Kwale oil-gas processing plant, a GGFR-supported project, uses associated gas for power generation at an independent power producer. The stated rationale for emissions reductions is that, in the absence of carbon finance, the returns to establishing the power plant would have been on the order of 13 to 15 percent, an inadequate inducement given the risky investment climate, including the risk of nonpayment by the electricity off-taker. (The data and assumptions underpinning this estimate were not made public.[4]) The estimated emissions reductions of 1.5 million tons CO_2e[5] will provide additional revenue, with security of payments guaranteed as long as the plant is running and producing electricity.

To jump-start the carbon market, the GGFR has supported studies, technical assistance, project preparation, and demonstration projects. One line of effort has been to develop and disseminate the methodologies needed to demonstrate emissions reductions. The Clean Development Mechanism (CDM) works by case law: once a methodology has been developed for a particular technology, subsequent similar projects can apply this methodology, substantially diminishing their development costs and reducing uncertainty about whether the project will be approved. The GGFR is developing two such methodologies for distinct approaches to flare reduction. The GGFR has also sponsored detailed screening exercises for Algeria and Indonesia to identify carbon investment opportunities. And it is involved in four CDM projects, two of which have been registered with the CDM, including the Kwale project, the largest CDM project in Africa. But uptake has been very slow: there are only 26 flaring-reduction projects in the CDM's pipeline of more than 3,000.

The program has devoted effort to developing carbon markets to help reduce flaring.

Divergent country outcomes defy easy generalization on the impacts of GGFR on flaring. In its early phases, the GGFR envisioned a goal of substantial absolute global reductions in flaring over 2005–2010 in a context of higher production.[6] Figure 6.2 shows total flaring by long-standing partner countries versus others over the past 12 years, using the remote sensing dataset. The graph compares the ratio of flaring in the two groups before and after the advent of the GGFR (2002); this provides a rough control for market changes that affect the production of oil or demand for gas.

Since that time, aggregate flaring by the GGFR partners has stayed roughly constant, while total flaring by nonpartners has increased. Much of the increase took place in Russia, outside partner region Khanty-Mansijsysk. Because flaring is related to oil production, it is interesting to track changes in flaring volume/barrels of oil produced. Cameroon's ratio was high and increasing until the advent of the GGFR, after which it declined. Four other GGFR members showed post-2002 declines in this ratio. In another three partners, a downward trend began before the GGFR and contin-

Flaring has stayed about the same among the GGFR partners and increased among nonpartners.

ued, while there was no trend or no change in four others. On balance, aggregate flaring by GGFR members stayed about the same in absolute terms but decreased relative to nonmembers, continuing a trend that was ongoing before the GGFR. Some flare reduction activities take years to implement. The expected release of remote sensing data for 2007 will provide an update on progress.[7]

Economics of Gas Flaring

Our review of GGFR studies and other analyses calls into question whether carbon markets address the root causes of much flaring, and whether carbon credits are necessary or sufficient to motivate flare reductions. The analysis for Indonesia (PA Consulting Group 2006), for instance, found that, in 10 of 26 fields analyzed, flared gas recovery projects offered substantial economic and financial returns even without carbon credits. If carbon finance were available at $15 per ton, it would boost the financial net present value of these potential projects by 6 to 13 percent. The report concludes that "the use of CDM adds value for project sponsors, but does not significantly change the

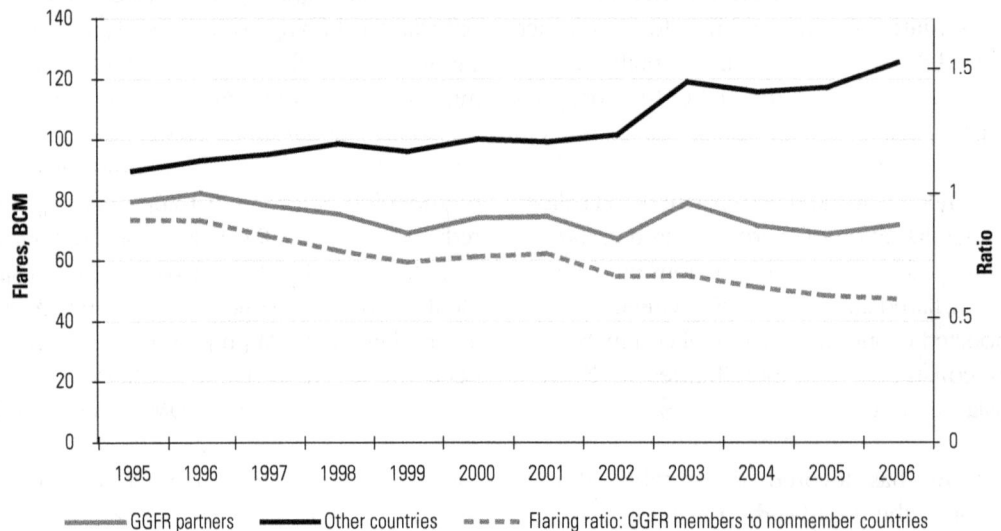

Figure 6.2: Global Flaring: Comparison of GGFR Partner and Nonpartner Countries

Source: IEG computations based on flaring data from http://www.ngdc.noaa.gov/dmsp/interest/ gas_flares.html (downloaded 9 June 2008), described by Elvidge and others 2007.

ranking of projects or make marginal projects highly attractive."

A reanalysis of the data by IEG shows that the economic rate of return (ERR) to associated gas use is large, even without carbon credits. When gas at one of these fields is used to generate electricity, and electricity is valued at long-run marginal cost, the ERR to capturing the otherwise wasted gas stream is an astounding 163 percent. That calculation assumes an oil price (the alternative fuel for power generation) of $70 per barrel. At $100 per barrel, the ERR soars to 223 percent. Even at $40 per barrel, the ERR is above 100 percent. In sum, economic fundamentals strongly support gas recovery in this case, even if global externalities are ignored.

However, financial rates of return, and the role of carbon, will be strongly affected by pricing policies. This is evident, for instance, in the IEA (2006) analysis of a typical potential flaring project in Russia. The internal rate of return (IRR) is −10 percent without carbon, but +5 percent with carbon, when gas is purchased at $22 per thousand cubic meters, which IEA viewed at the time as an institutionally determined price. But the returns rise to 23 percent (without carbon) and 32 percent (with) when the gas price is assumed to be $60—still substantially below the netback price that might be obtained if transmission to export markets were possible. Similar analyses were undertaken in a GGFR-commissioned study (PFC Energy 2007), suggesting that essentially all currently flared Russian gas could profitably be recovered at a gas price of $87, which is well below potential export values. Controlled prices may account for the low without-carbon internal rates of return that Kwale noted earlier.

Gas and energy pricing and regulatory policies are thus crucial considerations in increasing the availability of both associated and nonassociated gas. Policy issues stem from two dilemmas.

First, gas markets are not globally integrated with a market-determined price. Because gas transport is expensive, a gas producer may have just one potential buyer. The value to the buyer, at most, is the cost of using an alternative fuel or feedstock—for instance, the cost of fueling a generating plant with diesel rather than gas. This value may be considerably higher than the cost to the producer of capturing and transporting the gas. The gap between low supply price and high demand price represents economic rent, to be divvied up between buyer and seller, and can be a source of contention, especially when the true supply and demand prices are private information. (The same problem arises in determining the degree to which oil producers can afford to pay for flaring control out of oil profits.)

Carbon markets may not address the root causes of flaring and may not be necessary or sufficient to motivate flaring reductions.

The economic rate of return to the use of associated gas is high, but financial rates of return are strongly affected by pricing policies.

Second, the demand price may itself reflect distortions in downstream markets. When electricity or heat tariffs are kept artificially low, or when alternative fuels are subsidized, users' willingness to pay for gas is diminished. At economic prices for electricity, it would generally be economically and environmentally preferable to use associated gas for local electricity production rather than bear the costs (and incur the emissions) of transforming the gas to LNG for export.

Consequently, there can be tension among the goals of maximizing public revenues from gas exploitation, subsidizing the cost of downstream goods such as electricity and fertilizer, and providing adequate incentives and finance for extraction or recovery of gas. One danger is that regulators, not knowing producers' actual costs, may set gas prices too low to allow recovery or control of flaring, restricting the supply of gas (and possibly of oil as well). Another is that price controls, or restrictions on accessing export markets through pipelines or LNG, divert gas to lower-value or inefficient use, with the consequence that relatively clean and efficient sources of power are forgone. Policies with these outcomes could hurt the domestic economy while contributing to excessive CO_2 emissions. But full analysis of these policies would also entail looking at their distributional consequences.

The Bank has had long-standing engagement on gas policy (focusing on nonassociated gas) in some countries. Engagement was stronger in the 1980s when the Bank lent for gas development, declined as investment attention focused more on transmission and distribution networks, and may now be increasing. Analytic work, including analysis of the economic value of gas in alternative uses, has been a frequent feature of this engagement.

The impact of the Bank's engagement on gas reform is mixed, and attibution can be difficult.

The impact of the Bank's engagement in gas reform is mixed, and it can be difficult to attribute results to a given intervention. In Egypt, Bank engagement traces back at least to the early 1980s when it supported a number of gas investment projects. It continued through a 1990s investment project to a recently initiated project that seeks to promote use of LNG over heavily subsidized liquified petroleum gas. That engagement has had some positive outcomes. While gas has been subsidized, it has been explicitly subsidized at the consumer level rather than imposed through price caps on producer payments. Prices paid to producers (currently $2.65/mmbtu) are lower than the economic value (estimated at $3.65/mmbtu), but have still been sufficient to stimulate massive expansion in gas production and to switch Egypt's expanding power sector, at the margin, to gas from more polluting and carbon-intensive petroleum products. Recently announced reforms have boosted the price of gas to energy-intensive consuming industries from $1.10 to $2.65, which should encourage greater energy efficiency while reducing expenditures on subsidies.

The experience in Indonesia, which dates at least to pipeline projects of the early 1990s, has been less successful. Gas is purchased at low prices for many uses. For instance, although the potential netback price of LNG sales is $11 per thousand standard cubic feet, much gas is sold to petrochemical or fertilizer producers at $6. Gas transport policies may inhibit the ability of producers to find remunerative markets. One consequence is that although Indonesia flares about 3 bcm of gas per year, and much of that gas

would be readily recoverable at economic prices, 3,500 MW of gas turbines are being run on more polluting, more carbon-intensive, more expensive diesel (World Bank 2007e). And because electricity tariffs are held below the long-run marginal cost, the government is forced to subsidize the consumption of this diesel. However, Bank engagement continues, and the Bank has recently supported studies of gas pricing and pipeline policy.

Nigeria, the world's second-largest flarer, has reduced flares significantly over the past two decades through increased LNG exports, but it still has far to go to reach its long-standing goal of ending flaring in 2008. An ESMAP (2004) study outlined the scale of the problem: flaring consumes gas potentially worth $2.5 billion per year, while producing 70 million tons CO_2e of GHGs. The study found that prices of $0.75/mscf would be necessary to elicit supply of associated gas and $1.00/mscf for nonassociated gas. The study focused attention on supplying gas for domestic power generation, noting the importance of maintaining gas prices sufficient to elicit demand. A country review by IEG found little indication that the ESMAP study had been used until recently, and found inadequate attention by the Bank to these issues over the past eight years. However, the government of Nigeria announced a Gas Master Plan and pricing strategy in early 2008. The extent to which it draws on the ESMAP study or Bank advice is unclear.

Conclusion

While gas flaring is a complex phenomenon, economic fundamentals would often support the recovery and productive use of currently flared gas. Continued flaring thus reflects—in part—regulatory and policy failures, particularly in gas pricing. Where this is the case, the use of carbon finance as an instrument to reduce flaring is problematic. First, carbon payments may not change incentives significantly, even under current pricing policies. This would mean that such carbon projects are not additional, and that the carbon payments merely add to producer (or gas owner) profits. Second, policy or regulatory reform, though difficult, may offer greater and more widely shared domestic economic benefits. Finally, the option of carbon finance may reduce pressures for reform. Reforms, such as more effective enforcement of regulations against flaring and higher prices for associated gas, make recovery of associated gas more attractive, and thus undercut arguments for the additionality of carbon projects.

Carbon finance may nonetheless be justified for activities that are on the edge of economic viability, such as collection of associated gas from small sources and use for local poverty reduction.

Gas policy reform is not easy. National gas monopolies and other groups benefiting from the status quo may resist change. Where there is no financial need for Bank investment loans, opportunities for dialogue may be limited. Nonetheless, there are examples of success. World Bank experience shows that policy reform requires sustained engagement over long periods, detailed analytic underpinnings, and favorable political conditions. The GGFR can continue to contribute to this process by encouraging dialogue among stakeholders, by serving as an honest broker in discussions between governments and oil companies. Efforts to popularize the issue and to encourage independent monitoring of flare locations, volumes, and actors could help to create conditions for progress. These measures need to be complemented with continued cross-sector efforts, focused on the large flaring countries, to put flaring and gas policy into a broader cross-sectoral perspective.

Chapter 7

Wind turbines contrast with the architecture of the 300-year-old buildings of Bada Bagh, Rajasthan, India. Photo ©Jacqueline M. Koch/Corbis, reproduced by permission.

Findings and Recommendations

Over the years, the World Bank's strategic documents have pointed to three approaches to the promotion of climate mitigation activities that are consistent with developing countries' "common but differentiated responsibilities."

One approach involves assembling global funds to compensate nations for the added expense of undertaking low-carbon development projects. A second, related approach is to support technology research, development, and diffusion. These approaches are covered here only tangentially and will be a topic for the next phase of the climate evaluation. (See box 7.1 for a discussion of the challenges related to technology adoption.) A third approach is to pursue win-win or no-regrets policies and investments that offer both attractive domestic benefits and global gains.

Strategy documents dating to 1993 emphasize energy efficiency and removal of energy subsidies as important win-win approaches. This evaluation has mainly looked at policies in these two areas, which the IEA and others stress as key approaches to emissions reductions over the next 20 to 50 years. The evaluation has also discussed the specific issue of gas flaring, which can be seen as an example of both a pricing and an efficiency problem. Finally, the report has examined the potential trade-offs among growth, energy access for the poor, and emissions.

Findings

Development spurs emissions.

A 1 percent increase in income induces—on average, and with exceptions—a 1 percent increase in emissions. To the extent that the World Bank Group is successful in supporting broad-based growth, it will put pressure on climate change. This is the fundamental challenge of development in a carbon-constrained world and underlines the need to find countervailing strategies, especially for middle-income countries.

But there is no significant trade-off between climate change mitigation and energy access for the poorest.

The poorest people and the poorest countries currently emit only tiny amounts of GHGs, so growth for them puts no real pressure on the world's carbon budget. Basic electricity access for the world's unconnected households, under the most unfavorable assumptions, would add only a third of a percent to global GHG emissions, and much less if renewable energy and efficient light bulbs could be deployed. The welfare benefits of electricity access have been estimated in the range of $0.50 to $1 per kWh (IEG 2008e), while a stringent valuation of the corresponding carbon damages, in a worst-case scenario, is a few cents per kWh.

Policies can shape a low-carbon growth path.

The link between growth and emissions is strong but malleable. It is strong because income per capita and heating needs explain most of the 600-fold variation between countries in energy emissions per capita. It is malleable because there is still great potential for reductions. Although most countries follow a tight linkage of income to emissions, there is still a sevenfold variation between the most and the least emissions-intensive countries at a given income level.

Part of that variation is luck—including natural endowments of coal, gas, and hydropower—but it is also the product of policies that shape the use of those resources. So, in the relation between income per capita and emissions from power and heat generation, the share of electricity from hydropower accounts for half of the variation among countries not linked to income and heat needs.

Fuel pricing is a key policy affecting emissions.

This is especially clear for vehicle fuels, where high subsidizers—those whose diesel prices are less than half the world market rate—emit about twice as much per capita as other countries at similar income levels. Within the OECD, the countries that have maintained high fuel prices for decades (through taxation) have evolved more efficient transport systems. If all the member countries had maintained these levels, the OECD's emissions would be 36 percent lower (Sterner 2007).

Energy subsidies are large, burdensome, regressive, and damage the climate.

IEA (2006) estimates that energy subsidies outside the OECD cost a quarter-trillion dollars yearly. Subsidies also promote excessive GHG emissions. In many developing countries, these subsidies exceed the public expenditure on health, yet they are not well-targeted to the most vulnerable. Removal of these subsidies would bring domestic fiscal and economic dividends and could reduce global emissions by several percentage points.

The World Bank has been very active in supporting rationalization of energy pricing and increased collection.

The Bank has been a mainstay of power sector reform. While attribution is difficult, Bank-supported pricing reforms have often helped to boost tariffs and collection rates. Policy dialogue and analytic work have been associated with successful reforms. Success is noteworthy in many transition economies, which also recorded reductions in emissions per capita and emissions per dollar of GDP.

Country ownership of reforms is key. The prospect of EU accession has been a motivation for reform, and severe fiscal pressure has sometimes, but not always, facilitated reform. But tariff reform has been difficult where it threatens entrenched interests, such as agricultural users in India. Countries that are not under fiscal stress—such as those with ample oil revenues—are less likely to seek or accept World Bank advice on subsidy removal, especially with regard to implicit (off-budget) subsidies.

Although poorer groups often get a small share of energy subsidies, subsidy removal can threaten their welfare.

While some subsidies scarcely reach poor people, energy subsidies constitute 5 to 10 percent of household budgets of the lowest quintile in some countries. Removal of these subsidies can be painful to all, but especially dangerous for the poorest. Sharp increases in energy prices can be politically perilous, and are perceived as having sparked deadly riots and the fall of governments. The political feasibility of price rises, therefore, can depend on the presence of mechanisms that protect both vulnerable and influential groups.

One way to facilitate energy price adjustments is to couple them with social protection measures funded from the savings from reduced subsidies.

In Ghana, the government removed school fees and boosted funding of clinics in poor areas as compensation for gasoline and kerosene price rises. In Indonesia, an unconditional cash transfer, targeted to the bottom two income quintiles, was put in place to complement a steep fuel price rise. In both cases, ex ante analysis showed that lower-income groups would be better off, on average. In Armenia, a social transfer payment, designed to offset an electricity price hike, initially reached only 55 percent of poor people, but coverage is thought to have improved. But such compensatory programs may not be sufficient to secure the acquiescence of wealthier interests who benefit from subsidies.

Another potentially important way to ease the adjustment to higher energy prices is to couple price hikes with efficiency measures, so that net outlays on energy increase less steeply than prices.

In principle, subsidy savings could fund such efficiency investments. This adjustment technique has been little used to date (though see the discussion of the China Heat Reform Project, below). While several Bank projects promote mass distribution of compact fluorescent light bulbs, these have not been linked to tariff reforms.

End-user energy efficiency has been relatively neglected.

Efforts on energy efficiency, especially on the demand side, have been modest compared with its potential and its stated priorities. While country strategies for 20 of the 33 top emitters contained general references to energy efficiency, only 10 had specific objectives. About 5 percent of energy lending by volume since 1990 has been for components specifically related to

energy efficiency and district heating. (Efficiency gains may also accrue from improvements in transmission and distribution.) However, the limited evidence suggests that efficiency projects have had high rates of return compared with other energy sector projects, even without accounting for GHG benefits.

Policy engagement on efficiency has been even more limited.

Only 34 energy-efficiency projects supported by the World Bank during 1996–2007 included activities related to public policies, broadly construed. Among these, DSM projects have been limited in scope and sustainability because of a tendency to partner with utilities, which make money by selling electricity, and only in special cases by conserving it. Projects in standards and codes have succeeded in stimulating policy or regulatory change, but often devoted inadequate attention to institutions and implementation.

However, there have been some innovative efforts.

The China Heat Reform and Building Project is pursuing a difficult—but potentially very high pay-off—comprehensive policy and investment approach that promotes demand for and supply of efficiency. And there has been a spate of innovative projects in energy finance, including ESCOs, that seek to overcome credit market failures and transactions cost barriers. Although these contract-intensive institutions face challenges in weak institutional and legal environments, they appear to be expanding and will be assessed at greater length in Phase II of this evaluation. The availability of grant funds from GEF, ESMAP, and the Asia Sustainable and Alternative Energy Program has been critical in allowing staff to pursue innovative efficiency and renewables projects.

There are several reasons why energy-efficiency projects, and especially policy-oriented projects, appear to be underemphasized in Bank lending.

Internal Bank incentives work against these

93

projects because they are often small in scale and demanding of staff time and preparation funds. There is a general tendency (including among borrowers) to prefer investments in generation, which are highly visible and easily understood, to investments in efficiency, which are less visible, involve human behavior rather than electrical engineering, and whose efficacy is harder to measure. A neglect of rigorous monitoring and evaluation reinforces the negative view of efficiency. And investments often take place in the absence of an integrated resource plan (for power system expansion) that takes efficiency options into account. A paucity of Bank staff with expertise in efficiency (now being remedied) has both reflected and contributed to the neglect of the issue.

The Bank, through the GGFR, has fostered dialogue on gas flaring, but flaring activity has not yet been reduced.

Flaring of associated gas contributes more than 400 million tons of CO_2 to the atmosphere each year; if used for power, it would produce twice the amount consumed in Sub-Saharan Africa. Using flared gas is in many cases a clear win-win proposition. The Bank-hosted GGFR Partnership represents a modest but innovative effort to tackle this large problem. The GGFR has fostered dialogue on the issue among countries and oil companies, raised the issue's profile, and sponsored useful diagnostic analyses and data collection. The GGFR's global remote sensing survey of gas flaring provides objective and verifiable data in an area that is difficult to monitor and where some participants may have low incentives for accurate reporting. However, by 2006 there had not yet been any aggregate reduction in flaring among GGFR partners, although flaring per barrel of oil has decreased in most partner countries.

Carbon finance does not address the fundamental policy and institutional failures that cause gas flaring.

The GGFR has devoted attention to carbon finance as a means of flaring reduction, which is appropriate only where the economics of reduction are marginal. However, the GGFR's diagnostic work suggests that flaring often results from lose-lose policy-level natural gas pricing decisions rather than inherently marginal economics. Where this is so, the use of project-level carbon finance is a mere bandage for policy ailments that require a more fundamental cure.

Important information for the design and management of emissions-related policies is missing.

At the international level, there is no timely, comprehensive, and consistent monitoring of energy subsidies or prices. At the national level, there is a lack of basic data on key factors related to energy efficiency, such as technical losses in transmission and the extent and emissions of captive power plants. Lacking also are timely and accurate data on household, commercial, municipal, and industrial consumption and expenditures on energy. This makes it difficult to design and monitor the impact of price reform and efficiency policies. And monitoring and evaluation remain inadequate at the project level. For instance, only one of many compact fluorescent light distribution projects has built in a rigorous impact analysis.

The World Bank has a significant history of involvement with carbon accounting.

A pilot study on carbon shadow pricing was carried out 10 years ago, and carbon pricing is integral to the activities of the Bank's carbon funds. Carbon shadow pricing has been systematically incorporated in long-term planning of the expansion of Southeast Europe's power system. And the IFC has already adopted a Performance Standard that requires projects with significant GHG emissions to quantify them annually and to seek avenues for reducing them, including offsets. While there are important technical issues in footprinting and shadow pricing, these precedents suggest that they can be overcome and could be informative. However, quantifying the Bank's indirect and policy impacts on GHGs is more difficult, though these impacts may be larger than those of the direct, project-level effects.

Box 7.1: The Challenge of Catalyzing Technology Adoption

The next phase of the climate change evaluation will look in depth at the Bank Group's experience related to technology. The framework is presented here and may be helpful in exploring uses for the recently established Clean Technology Fund.

The public policy rationale for supporting renewable energy and energy efficiency revolves around barriers or market failures, including regulatory barriers, information and transactions costs, and spillover or demonstration effects that are not captured by innovators.

GEF climate projects and CDM projects are required to predicate their financial support for a project on a barrier-removal argument of this kind. IFC and IBRD/IDA support may do so implicitly. The next phase of the evaluation will examine a set of low-carbon technologies through this barrier-removal lens. Three evaluative questions stand out:

- Are the barriers as severe as they are represented to be? In the project context, could the project have been undertaken in the absence of concessional finance (known as the additionality test)?
- What are the spillover impacts of particular technology choices?
- What is the Bank's comparative advantage and how does that find expression in the strategic choices it makes among instruments and technologies?

On the *additionality* question, the experience of the CDM will be instructive, both for the Bank's expanded use of carbon finance (through the Carbon Prototype Fund) and for the deployment of the Clean Technology Fund. The CDM has built an elaborate apparatus to try to ensure additionality, project by project. Contentious from the start, the additionality tests are perceived as onerous red tape by some investors. At the same time, serious questions have been raised about whether these tests truly screen out projects that could have succeeded without carbon finance (Michaelowa and Purohit 2007; Schneider 2007; Wara 2008).

For instance, some observers cite a proliferation of CDM-financed hydropower plants in places where similar plants were already widespread. Analysis by the Bank's Carbon Finance Unit has shown that in many cases the sale of carbon offsets makes only a very small difference in the project's financial bottom line—a percentage point or less in the internal rate of return. For these projects, it is not plausible that carbon revenues alone were

enough to push the project over the threshold from unprofitable to profitable. However, the carbon finance transaction may have provided some other catalytic benefits. For instance, the due diligence associated with carbon finance may have crowded-in investors and financiers.

These additionality concerns are not unique to CDM projects. They also apply to pricing policies (such as feed-in tariffs or renewable portfolio standards) that promote renewable energy. As the level of support increases, to what extent is there a supply response, and to what extent do incumbents simply receive higher profits? This is a fundamental question to ask with regard to choosing mechanisms, technologies, and locations to support.

With regard to *spillover effects*, the technology projects with the most leverage are those that trigger spontaneous diffusion or replication. One well-known mechanism for spillovers is the learning curve. Technology costs decline with cumulative production volume, as has been well documented for solar photovoltaics and wind power. Taking advantage of these learning curves is inevitably an exercise in "picking winners," or at least short-listing them. Success is achieved when cumulative production of a particular technology is enough to push costs below the threshold of competitiveness.

Another mechanism is to reduce uncertainty among technology investors or users. For instance, the first wind or minihydro plant in a country or region may be viewed with skepticism. Risk-averse investors may demand a premium; lenders may simply be unwilling to lend. Successful demonstration of the technology in local circumstances could reduce the risk premium, making it easier for follow-on projects to get financing.

Finally, public policies can deter or enable investment. Subsidies to fossil fuels or red tape for small power producers are examples of deterrents. Building and appliance codes, in contrast, increase the salability of efficient building material and machinery.

A starting place for the discussion of the *Bank's comparative advantage* is to look at activities that are unattractive to the private sector, or the public sector in the developed world. Within that set, the Bank could focus on those that have the highest spillover effects.

These considerations suggest concentrating Clean Technology Fund and other new resources on technologies and activities that:

- Are not the subject of research and development in the developed world

(Box continues on the next page.)

95

Box 7.1: The Challenge of Catalyzing Technology Adoption (*continued*)

- Are easy to replicate and therefore difficult to protect by patent or other means
- Could be rapidly pushed down the learning curve
- Facilitate public sector activities that encourage investments in efficiency and renewables
- Cannot be financed through existing Bank instruments.

Examples include:

- Improved procedures for targeting social safety net payments to poor and vulnerable people, as a means to reduce energy subsidies
- Institutions, procedures, and technologies for ex ante assessment of energy consumption by buildings, and for implementing building code inspections
- Lower-cost technologies for delivering and installing (as opposed to manufacturing) efficiency measures and decentralized renewable power sources
- Low-cost technologies for DSM of traffic in high-density cities

- Detailed wind resource surveys for windpower site identification and investment decisions
- Geological surveys on the availability and integrity of carbon capture and storage sites
- Capacity building for regulators on integrated resource planning and on technology and regulatory issues for nuclear and carbon capture and storage technologies
- Land management techniques that reduce demand for energy-intensive fertilizer production
- Solar technologies of all kinds, given higher average insolation in developing regions.

Strategic consideration of these options will force some difficult choices. For instance, pursuing the learning curve route to technology commercialization requires focusing on a limited set of technologies and coordinating these investments across countries, while an emphasis on removing uncertainty as a barrier to investment would argue for a very diffuse set of investments across a wide range of countries.

Source: IEG.

Conclusion and Recommendations

The Bank is just one contributor toward the long-term goal of mitigating and adapting to climate change. The long-run solution to mitigation entails the invention and wide-scale deployment of zero-carbon technologies. Developing and deploying these technologies will require massive near-term increases in research and development expenditure in the developed countries, and trillions of dollars of investment—far beyond the Bank's direct financial resources—in developing countries.

Still, the Bank can aspire to play a catalytic role in this global transformation. Based on the analysis in this report, IEG makes the following recommendations.

Focus World Bank efforts more strategically on areas of its comparative advantage, which include supporting the provision of public goods and, at the country level, promoting policy and institutional reform.

The Bank has the potential to help its clients

pursue nationally appropriate actions that meet pressing development objectives, while positioning them on a lower-emissions growth path. It can best do so by seeking maximum leverage in its actions. This entails a strategic focus on its comparative advantage in supporting policy reforms, public goods, and institutional innovations that transform markets. There is ample scope for clients to pursue win-win policies—but if these were easy, they would have been undertaken long ago. Reform will require a systems view: looking at the power system as a whole; looking at energy subsidies as just one, dysfunctional, part of a social protection system; and looking at the connections between water and power management. And it will require big investments in real-time monitoring and learning.

Systematically promote the removal of energy subsidies, easing social and political economy concerns by providing technical assistance and policy advice to help reforming client countries find effective solutions, and analytical work demonstrat-

ing the cost and distributional impact of removal of such subsidies and of building effective, broad-based safety nets.

The mid-2008 level of energy prices, while burdensome for many countries, nonetheless prompts a fresh look at policies on energy subsidies, energy efficiency, and renewable energy sources. The recent experience of these prices may open doors for policy and regulatory reform. The Bank can provide analytic support for countries to explore the potential for gains from reform, and financial and technical support for carrying out reforms if desired.

Energy price reform is never easy or painless. It can endanger poor people, arouse the opposition of groups used to low prices, and trigger inflation, thereby posing political risks. But failure to reform can be worse, diverting public funds from investments that fight poverty and fostering an inefficient economy that is increasingly exposed to energy shocks. And reform need not be undertaken overnight. The Bank can provide assistance in charting and financing adjustment paths that are politically, socially, and environmentally sustainable.

One way to do this is for the Bank to continue to develop and share knowledge on the use of cash-transfer systems or other social protection programs as potentially superior alternatives to fuel subsidies in assisting the poor. To assist countries in dismantling subsidies that benefit special interest groups, the Bank should foster cross-sectoral cooperation and greater use of political economic analysis. Timely monitoring and analysis of energy use and expenditure, at the household and firm levels, will be important in policy design, in securing public support, and in detecting and repairing holes in the safety net.

Emphasize policies that induce improvement in energy efficiency as a way of reducing the burden of transition to market-based energy prices.

Cost-reflective prices for energy boost the returns to efficiency, but policies may need to be put in place to allow households and firms to exploit efficiency opportunities. Conversely, the deployment of energy-efficient equipment such as compact fluorescent lights can be used as a device for cushioning the impact of price increases. The Bank should explore innovative ways to finance efficiency (and renewable energy) investments in the face of fuel price volatility.

This report calls for much greater emphasis on promotion of energy efficiency. But similar calls in the past have not evoked a strong response. If a real reorientation to energy efficiency and renewable energy is to occur, the Bank's internal incentive system needs to be reshaped. Instead of targeting dollar growth in lending for energy efficiency (which may distort effort away from the high-leverage, low-cost interventions), it needs to find indicators that more directly reflect energy savings and harness them to country strategies and project decisions. It also needs to patiently support longer, more staff-intensive analysis and technical assistance activities. Increased funding for preparation, policy dialogue, analysis, and technical assistance is required. Trust fund resources have been helpful for this in the past; the Clean Technology Fund may provide additional, near-term funds.

Promote a systems approach by providing incentives to address climate change issues through cross-sectoral approaches and teams at the country level, and structured interaction between the Energy and Environment Sector Boards.

To tackle problems of climate change mitigation and adaptation, the Bank and its clients need to think beyond the facility level, beyond subsectors, and beyond sectors. The value of a windmill depends on the load patterns of the grid to which it is connected. Removing electricity subsidies for farmers requires an understanding of agricultural policies and conditions. Promoting municipal electricity efficiency is closely bound up with reducing distribution losses in water systems. Traffic congestion and air pollution are a consequence of fuel subsidies. Urban forestry promotes mitigation by cooling cities and fosters adaptation by reducing flooding.

To be effective, the Bank needs to break down sectoral stovepipes and encourage cross-sector approaches and teams. This will require championship by country directors and vice presidents. The unfulfilled promise of mainstreaming sustainable development needs to be realized through structured interaction of the Energy and Environment Sector Boards. This could be initiated with ad hoc groups to address specific cross-sectoral challenges.

At the country level, the Bank should support capacity building for a systems approach—for instance, for power system regulators in the area of integrated resource planning. And it should think about using the Clean Technology Fund to support public systems that will catalyze widespread investments. For instance, capacity building for building inspectors and for the construction industry could transform that industry.

Invest more in improving metrics and monitoring for motivation and learning—at the global, country, and project levels.

Good information can motivate and guide action. Building on the Bank's current collaboration with the IEA or other partners on energy-efficiency indicators, the Bank should set up an Energy Scoreboard that will regularly compile up-to-date standardized information on energy prices, collection rates, subsidies, policies, and performance data at the national, subnational, and project levels. Indicators could be used by borrowers for benchmarking; in the design and implementation of country strategies, including sectoral and cross-sectoral policies; and in assessing Bank performance. The Bank could look for inspiration to India, which already publishes detailed data on power plant CO_2 emissions, state-level utility performance, and fuel subsidy levels, or to China, which is aggressively pursuing a goal of energy-efficiency improvement.

At the national level, the Bank should support integration of household and firm surveys with energy consumption and access information to lay the foundation for assessing impacts of price rises and mitigatory measures, as well as planning for improved access. The Bank could explore the use of advances in information technology (such as meters with automated, wireless reporting), together with statistical sampling, to undertake real-time monitoring of energy use and patterns. Affordable monitoring systems could pay big dividends in improved energy management at the sectoral and national levels.

More rigorous economic and environmental assessment is needed for energy investments and those which release or prevent carbon emissions. These assessments should draw on energy prices collected for the Energy Scoreboard and account for price volatility. In addition, they could undertake carbon accounting at the project level, computing switching values for high- and low-carbon alternatives. Investment projects should also be assessed, qualitatively, on a diffusion index, which would indicate the expected catalytic effect of the investment on subsequent similar projects. Where proprietary information is not involved, these assessments should be made public for information and comment. Public disclosure will provide incentives for accurate assessment and will also inform global technology and investment planning.

Ideally, investments should fund projects identified under an active integrated resource plan for system expansion. Such plans should allow for energy efficiency as a source of increased capacity and take account of the value of renewables in reducing pollution and exposure to external price shocks. The Bank should assist countries in preparing and implementing these plans. Countries may wish to compare expansion plans under different shadow prices for carbon.

It is desirable to complement project-based analysis with assessment of indirect and policy-related impacts, which could be much larger.

Monitoring and evaluation of energy efficiency interventions continues to need more attention. Large-scale distribution of compact fluorescent light bulbs is one example of an intervention that is well suited to impact analysis and where a timely analysis could be important in informing possibly massive scale-up activities.

Appendixes

| Country | Total subsidies | Subsidy | | | PER attention |
		Electricity	Fuel	Gas	
Argentina	1990–91: $0.6 bln[a] 1995–96: $0.15 bln[a]	$1.5 bln[b]	$1.0 bln[b]	$4.0 bln[b]	2003 +
China	1990–91: $24.5 bln[a] 1995–96: $10.3 bln[a]	$5.0 bln [b]	$7.0 bln[b]	$4.0 bln[b]	No PER
Egypt, Arab Rep. of	11.9% of GDP[c]	$2.0 bln[b]	$9.5 bln[b]	$1.0 bln[b]	No PER
Indonesia	1990–91: $2 bln[a] 1995–96: $1.3 bln[a] PER: 2006–$12 bln	$2.0 bln[b] PER $3.8 bln. =1.4% of GDP, includes explicit and implicit (2005)	$15.5 bln[b] PER: 1.5% GDP 2004–3% 2005–3.5%		2007 ++
India	1990–91: $4.2 bln[a] 1995–96: $2.7 bln[a]	$10.0 bln[b]	$7.0 bln[b]	$2.0 bln[b]	No PER
Iran, Islamic Rep. of	17.5% of GDP[c]	$2.5 bln[b]	$24.0 bln[b]	$9.5 bln[b]	No PER

CAS objectives related to energy pricing or subsidies	Outcomes and status
2004: Social tariff to be targeted to the poor. Sanction temporary seasonal price adjustment mechanisms in the energy sector.	**2006:** Social tariff: draft law before Congress, but government is unreceptive. **2007:** Electricity tariffs raised but still substantially below long-run marginal cost; gas subsidies in place.
1995: Average power tariffs should approach long-run marginal cost countrywide, but especially in the interior provinces. **1997:** Enact price reforms and user fees to increase retained earnings of infrastructure companies. No energy pricing/subsidy target in CASs.	**2003:** Average power prices cover cost; higher than long-run marginal cost in coastal provinces. **2007:** GoE announced plans to eliminate gas and electricity subsidies for energy-intensive industries over the coming three years. Bank provided input on energy prices and subsidies. **2008:** Increases in electricity, natural gas, and fuel prices, but still low by international and regional standards. The Ministry of Finance started to record energy subsidies in the budget in 2005/06 to increase transparency.
1995: Institute electricity rate increases. **1997:** Raise domestic fuel prices. Power: increase household electrification ratio; raise electricity tariffs. **2001:** Phase out fuel and power subsidies to solve the problem with power sector bottlenecks. **2004:** Cost-effective tariff/user charges policies; automatic tariff adjustment mechanisms for power.	**1999:** Financial crisis constrains tariff hikes. **2003:** Some tariff increases in electricity, but prices still inadequate to attract investors. **2005:** Fuel price hiked, with compensatory targeted assistance to the poor. An unconditional cash transfer program reached 19.2 million poor and near-poor households (34% of the national population). Fuel price adjustments saved $15 bln in public funds over 2005–06. **2006:** Electricity and petroleum prices remain below cost levels as world prices rise.
1995: Depoliticize tariff adjustments and other decisions in power sector. **1998:** Implement power tariff adjustments. Liberalize coal pricing and distribution. **2002:** Bring down theft and losses. Reduce fiscal drain of the power sector. Better cost recovery for power through appropriate tariff schedules, lower subsidies, and reversal in culture of nonpayment. **2005:** Tariffs should cover the cost of service provision. Progressively reduce the primary deficit at the center and in states by reducing power sector losses and phasing out petroleum subsidies.	**1999:** Adjustment of domestic diesel prices (40% price hike). Higher electricity tariffs, reduced power subsidies. **1996-2006:** Bank's five state-level power restructuring projects fail to achieve expected tariff increases. **2008:** After-tax petroleum prices are above world market levels; kerosene heavily subsidized.
2001: Interim Assistance Strategy Country Economic Reform agenda addresses subsidies. Bank will intensify economic and sector work, including a study on the reform of the energy pricing system.	**2001:** Study on the reform of the energy pricing system finalized. Subsidies continue.

(Continues on the next page.)

Bank Attention to Subsidies in the Large Subsidizing Countries (*continued*)

Country	Total subsidies	Subsidy			PER attention
		Electricity	Fuel	Gas	
Kazakhstan		$0.5 bln[b]	$1.0 bln[b]	$4.0 bln[b]	No PER
Malaysia		$0.4 bln[b]	$3.5 bln[b]	—	1999 +
Nigeria	1990–91: $0.9 bln[a] 1995–96: $0.5 bln[a] PER: power and steel sectors (% of total budget): 1998—2.6% 2000—7.9% Implicit fuel subsidies: 2003—1.6% of GDP[c] 2005 estimate: 2.2% of GDP[c]	$0.4 bln[b]	$2.0 bln[b]		2001 +
Pakistan	Explicit fuel subsidies: 2003—0.1% of GDP[c] 2005 estimated: 0.2% of GDP[c]		$2.0 bln[b]	$3.0 bln[b]	—
Russian Federation	2003 (PER): 3.3% of GDP in housing and utility subsidies, direct budget support, quasi-fiscal financing	$14.0 bln[b]		$26.0 bln[b]	2005 ++

CAS objectives related to energy pricing or subsidies	Outcomes and status
1998: Liberalization of pricing and regulations in energy. **2002:** Government will introduce a new tariff policy that will fully cover the cost and will be aimed at reducing the independence of the tariff on electricity transmission distance. No energy pricing/subsidy target in CAS. No energy pricing/subsidy target in CAS.	**2000:** Government implemented the replacement of the petroleum subsidy by a consumption tax resulting in a doubling of the gasoline price. **2004:** Nigeria successfully implemented a fiscal rule de-linking the budget from current oil prices.
1995: Improve structure of energy prices. Increase gas, petroleum, and electricity prices. Introduce an automatic adjustment mechanism for petroleum prices. **1998:** Increase electricity tariffs, with possible delay in petroleum price adjustment. **2002:** Increase in electricity tariffs, gas prices (to be linked to international crude price), with possible delay in petroleum prices adjustment. **2006:** Oil and gas prices should fully reflect world market conditions.	**1998:** Electricity tariffs increased by 21%. **2001:** Formula-based adjustment of petroleum prices introduced. Consumer gas prices increased, but gas is still sold at $0.77/mmbtu when opportunity cost is $1.76/mmbtu. **2004:** Gas subsidy to fertilizer industry continued, electricity rates were reduced despite continued high losses. Little progress in reducing public sector arrears. Petroleum prices adjusted partially to reflect international oil prices. **2005:** Government temporarily suspended policy of automatic petroleum price adjustment. Reduction in petroleum taxes, to reduce the impact of rising international prices. Gas tariff collections met policy goals but price adjustment mechanism is not consistent. Gas tariffs are priced close to long-run costs on average but are still below opportunity costs for households and for the fertilizer industry. **2006** Inability of government to adjust petroleum consumer prices as foreseen in the CAS. Good progress made in the pricing of natural gas, but gas tariffs continue to be distorted by cross-subsidies among different classes of consumer, and implementation of the gas price adjustment mechanism has been erratic.
1995: Pricing, cost recovery in utility sector (power). **1997:** Reduction in subsidies to coal enterprises for investment and production. **1999:** Energy sector—district heating—tariff level; structure; cash collection. Coal—improve subsidy management system, social safety net, level/composition of subsidies and sector governance. Indicator: Real reduction in level of coal sector production subsidies. **2002:** Reduction of subsidies in energy sector (coal, district heating, power tariffs).	Substantial improvements in cash collection. **1997:** Introduced new pricing principles in infrastructure monopolies—electricity, natural gas, and railways—to cost-based pricing and reduced cross-subsidization. Electricity prices to industrial customers are comparable to many OECD countries. **1999:** State subsidies to the coal industry declined from 1% to 0.2% of GDP. **2002:** Substantial improvements in cash collection in electricity and gas. **2005:** Subsidies to natural monopolies reduced. Though tariffs are being gradually increased toward long-run marginal costs, price subsidies remain substantial. Domestic prices of gas still well below export levels.

(Continues on the next page.)

Bank Attention to Subsidies in the Large Subsidizing Countries (*continued*)

Country	Total subsidies	Subsidy			PER attention
		Electricity	Fuel	Gas	
South Africa	1990–91: $0.9 bln[a] 1995–96: $0.4 bln[a]	$4.0 bln[b]			No PER
Thailand	1990–91: $0.5 bln[a] 1995–96: $0.4 bln[a]	$1.0 bln[b]	$2.0 bln[b]	$0.3 bln[b]	No PER
Ukraine	PER: Quasi-fiscal activities in the energy sector (% GDP): 2001: 7.4; 2005: 4.3	$2.5 bln[b]	$0.3 bln[b]	$13.0 bln[b]	2006 ++
Venezuela, R. B. de	1990–91: $3.4 bln[a] 1995–96: $2.4 bln[a]	$1.5 bln[b]	$8.5 bln[b]		No PER
Vietnam		$0.7 bln[b]	$0.7 bln[b]		No PER
Mexico	1990–91: $5.4 bln[a] 1995–96: $2.2 bln[a]	PER: 1% of GDP in 2003			2005 ++

Sources for subsidy estimates:

a. World Bank 1995.

b. Estimate of energy subsidies based on IEA 2007, figure 11.7: Economic Value of Energy Subsidies in non-OECD Countries for 2005.

c. Baig and others 2006.

d. Bacon and Kojima 2006.

Note: No PER = no PER implemented; – = no subsidy analysis in PER; + = perfunctory analysis and general recommendations; ++ = detailed analysis and specific recommendations.

CAS objectives related to energy pricing or subsidies	Outcomes and status
2007: Assess competition and regulation in the energy sector.	
1995: Assessment of impact of power tariff and connection charges.	Fuel subsidies phased out over 2004–05.
2000: Improve financial discipline across the economy (including energy sector—full payment in cash). **2003:** Financially sustainable sectors: Energy and Infrastructure (tariff, regulatory, and old debt issues addressed). Set tariffs sets close to cost recovery (coal, gas, electricity). **1997:** Maintain domestic petroleum prices at export parity.	**1999–2006:** Collection rates in the energy sector increased from 8% to 98%, coal tariffs doubled, gas tariffs increased by 25%, and electricity by 47%. CO_2 emissions/$ dropped substantially. Bank lending and advisory work helped. However, the rise in energy prices leaves tariffs still below economic costs.
1996–2002: Raise power tariffs to long-run marginal cost countrywide. **2003:** Rationalize pricing policies for infrastructure policies. **2007:** Institute cost-effective electricity tariffs for different consumer categories.	**2002:** Tariffs are still below long-run marginal cost. **2006:** Gradual convergence to regional prices. Cost recovery in electricity.
1997: Reduce price distortions in infrastructure, technical assistance on tariff policies in power sector. **1999:** New approach to pricing and subsidization needed at both the national and subnational levels. **1997:** "On the top of environmental agenda is efficient energy pricing." **2002:** "Revise pricing policies and subsidies (including energy) that claim to assist the poor, actually convey perverse signals and induce overuse, misallocation, and waste of environmental assets." **2004:** Better targeting of subsidies. **2008:** Electricity subsidies study.	

Fiscal year	Project ID	Project and status (closed [C]/active [A])		IBRD/IDA grant amount (US$ million)	IEG outcome rating	Type of efficiency measures
1996	P034491	Albania Power Transmission and Distribution—IBRD/IDA	C	29.50	Unsatisfactory	Appliances and buildings standards
1996	P034617	Mali Selingue Power Rehabilitation Project—IBRD/IDA	C	27.30	Highly satisfactory	DSM study
1997	P035693	China Efficient Industrial Boilers—GEF	C	32.80	Satisfactory	Technology diffusion
1997	P035163	Lithuania Energy Efficiency/ Housing Pilot Project—IBRD/IDA	C	10.00	Moderately satisfactory	Efficiency finance fund
1997	P042056	Senegal SN-GEF Energy Mgmt Sust Prtn SIL—GEF	C	4.70	Not rated	Technology diffusion
1997	P010498	Sri Lanka Energy Services Delivery—IBRD/IDA	C	24.20	Satisfactory	Voluntary building standards; capacity building for DSM
1998	P000532	Chad Household Energy Project—IBRD/IDA	C	5.30	Moderately satisfactory	Household energy DSM/ technology diffusion
1998	P003606	China Energy Conservation— IBRD/IDA	C	63.00	Satisfactory	ESCO demonstration and market information
1998	P037859	China Energy Conservation— GEF	C	22.00	Satisfactory	Energy conservation information center
1998	P000736	Ethiopia Energy 2—IBRD/IDA	C	200.00	Moderately satisfactory	Electric Standards and Technical Regulation study

Projects with Energy-Efficiency Policy Component, 1996–2007

Energy-efficiency component, (US$ million)	Efficiency measures detail	Outcome
8.7 (metering)	Purchase of meters and related accessories. UNDP-financed studies on appliance efficiency and energy conservation in buildings.	Meters purchased. The study on energy conservation in buildings was completed and a law was passed in 2002 to prescribe insulation standards. The study on appliance efficiency was not done.
Not identified	Purchase and installation of equipment for reducing distribution network technical losses and the design and promotion of end-use efficiency programs.	Never implemented
31.49	Development, production, marketing of energy-efficient and cleaner industrial boiler designs.	Achieved
9.80	Loan finance for energy-efficient rehabilitation of residential buildings.	The investments introduced controllable heat consumption.
0.33	Demand management and fuel substitution component to promote substitution of kerosene and liquid petroleum gas for charcoal, and will disseminate efficient charcoal stoves.	Achieved
1.9	Energy-efficiency objectives within the capacity-building component.	The project launched DSM programs, including: a code of practice for energy-efficient commercial buildings; increased technical capacity to carry out energy audits and provide advice on energy efficiency measures; and an appliance energy-labeling program.
0.53 actual	Improve the efficiency of household fuel use: DSM to reduce wood fuel consumption, through: (a) commercialization of efficient cooking stoves (firewood, charcoal); and (b) promotion of the use of low-cost kerosene and liquefied petroleum gas stoves.	Not achieved, subcomponent was discontinued—production problems and lack of demand.
57.96	Adapting the EPC (energy performance contract) model; energy management company demonstration; information dissemination.	Energy management company demonstration. By 2007, the energy efficiency achieved total energy savings of 5.92 mmtce, and associated reductions in carbon dioxide emissions of 5.06 million tons of carbon equivalent versus the target of 3.77 in 2002.
22	The Energy Conservation Information Dissemination Center (NECIDC).	NECIDC was established.
Not identified	Supply-side efficiency investment—energy-efficiency subcomponent related to increasing efficiency in the power sector.	Improve utilization efficiency of rural renewable energy: not achieved.

(Continues on the next page.)

Projects with Energy-Efficiency Policy Component, 1996–2007 (*continued*)

Fiscal year	Project ID	Project and status (closed [C]/active [A])		IBRD/IDA grant amount (US$ million)	IEG outcome rating	Type of efficiency measures
2000	P047309	Brazil Energy Efficiency—GEF	C	15.00	Moderately satisfactory	Standards, testing, capacity building
2000	P066345	Mauritania Energy, Water, and Sanitation Sector Reform Technical Assistance Project—IBRD/IDA	C	9.90	Unsatisfactory	Energy-saving action plan
2002	P074040	Bangladesh Renewable Energy Development—GEF	A	8.20	Not rated	DSM and master plan
2002	P063644	Ecuador Power and Communications Sectors Modernization and Rural Services— IBRD/IDA	A	23.00	Not rated	Standards, tariffs incentives for energy conservation in electricity; efficiency finance
2002	P072527	Ecuador Power and Communications Sectors Modernization and Rural Services—GEF	A	2.80	Not rated	Building standards and efficiency
2002	P076702	Sri Lanka Renewable Energy for Rural Economic Development—IBRD/IDA	A	75.00	Not rated	DSM technical assistance, including efficiency finance
2002	P066396	Vietnam System Efficiency Improvement, Equitization, and Renewables Project—IBRD/IDA	A	225.00	Not rated	Utility-based DSM
2003	P076977	Brazil Energy Sector Technical Assistance Project—IBRD/IDA	A	12.10	Not rated	Expansion planning, tariff reform with targeted smart subsidies
2003	P049395	Ethiopia Energy Access SIL—IBRD/IDA	A	132.70	Not rated	Technology diffusion

Energy-efficiency component, (US$ million)	Efficiency measures detail	Outcome
11.9	Capacity building for improving the efficiency of electricity use by the residential and commercial sectors in Brazil. Removal of barriers to energy efficiency and energy conservation.	GEF project implemented, IBRD loan canceled. Energy Efficiency Reference Center is fully operational. Marketing Plan and Publicity Campaign implemented. National Electric Laboratory is fully operative—testing and labeling programs in place. Testing, Certification, and Labeling Program supported the implementation of the Law on Energy Efficiency (approved October 2001). Research on acceptance of efficiency equipment by market. Training on energy-efficiency management and capacity-building program.
Not identified	The preparation of an energy-saving action plan for public services.	Studies deemed of limited utility. Complementary tariff reforms not enacted.
Not identified	Introducing standards and programs for testing and certification.	
1.74	Develop strategies and policies to remove barriers; standards for efficient design and use of buildings and electrical appliances, public information, and support to the establishment of ESCOs.	
Not identified	Design of energy-efficiency standards for buildings and energy equipment.	
0.75	Energy efficiency and DSM. Technical assistance and credit support for provision of energy-efficiency services. Public policy and ESCO elements.	
5.5	DSM components.	
1.4	Increasing access to and affordability of electricity, natural gas, and liquefied petroleum gas including through tariff reform with targeted, smart subsidies. Demand- and supply-side possibilities—through training and capacity building.	
Not identified	Promotion of commercially based production and dissemination of approximately 320,000 improved baking stoves.	

(Continues on the next page.)

Projects with Energy-Efficiency Policy Component, 1996–2007 (*continued*)

Fiscal year	Project ID	Project and status (closed [C]/active [A])		IBRD/IDA grant amount (US$ million)	IEG outcome rating	Type of efficiency measures
2003	P071019	Vietnam Demand-Side Management and Energy—GEF	A	5.50	Not rated	Utility-based DSM
2004	P073036	Mali Household Energy and Universal Access Project—IBRD/IDA	A	35.70	Not rated	Energy services, charcoal-efficiency promotion
2004	P074686	Morocco Energy and Environment Upgrading—GEF Medium Size	C	0.80	Not rated	Capacity building for efficiency finance
2004	P068124	Uruguay Energy Efficiency Project—GEF	A	6.90	Not rated	Utility-based efficiency services
2005	P077575	Bulgaria District Heating—Carbon Offset	A	4.30	Not rated	DSM public awareness
2005	P069126	Burkina Faso Power Sector Development—IBRD/IDA	A	63.60	Not rated	DSM institutional strengthening
2005	P072721	China Heat Reform and Building Energy Efficiency Project—GEF	A	18.00	Not rated	Codes and standards; metering
2005	P070246	Poland Energy Efficiency—GEF	A	11.00	Not rated	Efficiency finance, capacity building, and demonstration

Energy-efficiency component, (US$ million)	Efficiency measures detail	Outcome
5.5	1. Electricity of Vietnam's DSM Program: expanded time-of-use metering, pilot direct-load control program, CFL promotion, fluorescent tube lamp market transformation; 2. Ministry of Industry Pilot Commercial Energy Efficiency Program: training, subproject financing, grants, program marketing.	
10.56	Interfuel substitution and household energy efficiency.	
Not identified	Energy (electricity and fuel) savings in an industrial park in Casablanca by facilitating local ESCO businesses and strengthening the local executing agency.	
6.48	Improved efficiency of energy use: energy-efficiency market development (implemented by the Ministry of Industry, Energy, and Mining); utility-based energy-efficiency services.	
4.34	Rehabilitation of pipeline and district heating substations. The technical assistance supported by the KIDS component of the project will support a public awareness campaign to promote DSM.	
3.38	1. Design a DSM policy framework and implement DSM programs, including capacity building in Energy Management Unit, to build a base for building code and appliance standards. 2. Purchase of efficient air-conditioning and lighting system. 3. Information awareness campaign.	
18	Promote simultaneous development of both heating sector reforms and building energy-efficiency improvements in 4-6 northern Chinese municipalities, achieving broad national impact.	
12.37	Components: 1. A partial guarantee. 2. Investments in bundled energy efficiency projects in the Krakow region. 3. Technical assistance for: deployment of guarantee mechanism; ESCO subsidiary in the development of the performance contracting model; training to local banks; awareness and demand for efficiency investments; project monitoring data and dissemination of results.	

(Continues on the next page.)

Projects with Energy-Efficiency Policy Component, 1996–2007 (*continued*)

Fiscal year	Project ID	Project and status (closed [C]/active [A])		IBRD/IDA grant amount (US$ million)	IEG outcome rating	Type of efficiency measures
2006	P086379	Djibouti Power Access and Diversification—IBRD/IDA	A	7.00	Not rated	Studies on sector efficiency
2007	P097635	Kosovo Lignite Power TA—IBRD/IDA	A	8.50	Not rated	Technical assistance policies and strategies on energy efficiency
2007	P099618	Morocco Energy Sector DPL—IBRD/IDA	C	100.00	Not rated	Standards, energy-efficiency audits, public policy, efficiency finance
1996	P091074	Philippines Public and Private Sectors Capacity Building Project (TF028553)—IDF	C		Not rated	Capacity building for DSM policy development
1997	P039965	Sri Lanka Energy Services Delivery—GEF	C	5.90	Not rated	DSM policy; efficiency finance
2004	P066532	Philippines Electric Cooperative System Loss Reduction Project—GEF	A	12.00	Not rated	Capacity building for efficiency finance

Source: IEG based on project appraisals, Implementation Completion Reports, Implementation Completion Report reviews, and Project Performance Assessment Reports.

Energy-efficiency component, (US$ million)	Efficiency measures detail	Outcome
< 0.5	Technical assistance directed on tariffs and losses study.	
Not identified	Policies and strategies to promote renewable energy, cogeneration, and energy efficiency in Kosovo.	
16	1. Set energy-efficiency standards for appliances, appliance labeling, new buildings, street lighting, and public buildings. 2. Organize the execution and monitoring of mandatory energy-efficiency audits in large and medium-size industries. 3. Government support for financing of energy efficiency programs. Conditionality: Adoption of implementing decrees on Energy Efficiency and Renewable Energy.	
Not identified	Support to the process of public consultation to design a DSM strategic plan and foster ownership of DSM programs through shared energy efficiency and energy conservation objectives.	
1.06	Strengthen the environment for DSM implementation, and improve the public and private sector performance to deliver energy services through renewable energy and DSM.	
12.00	Partial Credit Guarantee Program. Capacity building: 1. Provide technical assistance and training to financial intermediaries, electric cooperatives (ECs), and ECs' investors, in power distribution systems; 2. Technical assistance to the Department of Energy for energy-efficiency gains of ECs from improved access to commercial lending.	

Country	Energy subsector	Source	Energy expenditure as a share of household expenditure	
			Poor	Non-poor
Albania	No subsidy	PSIA, 2003	>20% on electricity of the cash income of the core poor (bottom 12%)	
Argentina	Patagonian gas	PER, Sept. 2003	...	...
Armenia	Utilities	PSIA, 2001	18% on utilities such as telephone, gas, central heat, and water for the poor	11% on utilities such as telephone, gas, central heat, and water for the non-poor
	Electricity	Tajikistan PSIA, 2006	13.9% for bottom quintile	6.5% for top quintile
Bangladesh	Residential gas	PER, Sept. 2003	...	...
Bolivia	Hydrocarbon derivatives	PSIA 2004, Coady, Grosh, and Hoddinott 2006	7.8% on transportation, fuel, and lubricants	9.3% on transportation, fuel, and lubricants
	Transportation	PSIA, 2004	5.4% on transportation	5.9% on transportation
	Fuels and lubricants for vehicles	PSIA, 2004	0.6% on fuels for vehicles	2.2% on fuels for vehicles
	Fuels and lubricants to cook	PSIA, 2004	1.9% on fuels for cooking	1.2% on fuels for cooking
	LPG	Coady, Grosh, and Hoddinott 2006	2.6% for bottom quintile	1.1% for top quintile
	Gasoline and diesel	Coady, Grosh, and Hoddinott 2006	0.0% for bottom quintile	2.5% for top quintile
Djibouti	No subsidy	PSIA, 2005	18.2% (18.5)[a] on biomass, electricity, and other energy sources in bottom quintile	16.3% (13.4)[a] on biomass, electricity, and other energy sources in top quintile
Ecuador	Cooking gas	PER, Nov. 2004	...	...
	Electricity, cooking gas and fuel	PER, Nov. 2004	...	...
Egypt, Arab Rep. of	Kerosene	PSIA, 2005	...	...
	Natural gas	PSIA, 2005	...	...

Income impact of subsidy removal/price increase		Share of benefit from the energy subsidy	
Poor	Non-poor	Poor	Non-poor
...	...	...	...
...	...	Subsidy excludes 95% of poor gas consumers, while 63% of its beneficiaries are non-poor; 100% of the resources of the subsidy go to the far south, which has only 3% of the nation's poor.	
9 percentage point increase in expenditure for the poor	3 percentage point increase in expenditure for the non-poor	...	...
...	...	...	...
...	...	The 4% of households with gas access receive Tk1.6 billion in implicit subsidies	
5.8% reduction in real income for bottom quintile	4.7% reduction in real income for top quintile	15.3% for bottom two deciles	...
...	...	...	...
...	...	...	...
...	...	...	...
...	...	...	...
...	...	...	...
...	...	...	...
...	...	3% to bottom quintile	17% to top quintile
...	...	...	67% to top four deciles
2.2 percentage point reduction of income for bottom quintile	0.1 percentage point reduction of income for top quintile	...	...
0.1 percentage point reduction of income for bottom quintile	0.6 percentage point reduction of income for top quintile	...	...

(Continues on the next page.)

Distributional Incidence of Subsidies (*continued*)

Country	Energy subsector	Source	Energy expenditure as a share of household expenditure	
			Poor	Non-poor
Egypt, Arab Rep. of (continued)	Gasoline	PSIA, 2005	...	...
	LPG	PSIA, 2005	...	...
	Above four products	PSIA, 2005	...	...
Georgia	Electricity	Tajikistan PSIA, 2006	6.3% for bottom quintile	2.0% for top quintile
Ghana	Petrol	Coady and Newhouse 2006	0.1% for bottom quintile	2.1% for top quintile
	Kerosene	Coady and Newhouse 2006	5.9% for bottom quintile	1.6% for top quintile
	LPG	Coady and Newhouse 2006	0.0% for bottom quintile	0.2% for top quintile
	Above three products	Coady and Newhouse 2006; Coady and others 2006	...	...
	Electricity	PSIA, 2004	...	...
Hungary	Electricity	Tajikistan PSIA, 2006	6.5% for bottom quintile	3.7% for top quintile
Indonesia	Fuel	DPL-II Program Document, 2005	3.6% for the bottom two deciles spent on fuel (3% on kerosene)	...
	Electricity	PER, 2007	...	...
Jordan	Fuel	Coady and others 2006	7.1% for bottom quintile	7.1% for top quintile
	Kerosene	Coady and others 2006	1.0% for bottom quintile	0.3% for top quintile
	LPG	Coady and others 2006	1.8% for bottom quintile	0.7% for top quintile
	Gas, regular	Coady and others 2006	0.9% for bottom quintile	2.3% for top quintile
	Gas, premium	Coady and others 2006	0.0% for bottom quintile	1.1% for top quintile
	Diesel	Coady and others 2006	0.3% for bottom quintile	0.9% for top quintile
	Electricity	Coady and others 2006	3.1% for bottom quintile	1.8% for top quintile
Kazakhstan	Electricity	Tajikistan PSIA, 2006	0.9% for bottom quintile	0.6% for top quintile
Mali	Fuel	Coady and others 2006	...	...

Income impact of subsidy removal/price increase		Share of benefit from the energy subsidy	
Poor	**Non-poor**	**Poor**	**Non-poor**
0.04 percentage point reduction of income for bottom quintile	1.4 percentage point reduction of income for top quintile	...	...
5.4 percentage point reduction of income for bottom quintile	2.0 percentage point reduction of income for top quintile	...	...
7.7 percentage point reduction of income for bottom quintile	4.1 percentage point reduction of income for top quintile	13% for LPG, kerosene, gasoline, and natural gas for bottom quintile	34% for LPG, kerosene, gasoline, and natural gas for top quintile
...	...	...	...
Progressive	...	...	
Regressive	17.8% for bottom quintile	20.9% for top quintile	
...	...	...	...
9.1% reduction in income	8.2% reduction in income	23.0% for bottom four deciles	...
Lifeline tariff is 4% of income for the lowest decile of connected households	...	...	...
...	...	...	...
The fuel price hike corresponded to 5.1% of per capita expenditure for the bottom decile	The fuel price hike corresponded to 6.2% of per capita expenditure for the top decile	The subsidies accruing to the top decile from fuel subsidies were 5 times those accruing to the bottom decile. The top 40% got 60% of the subsidy.	
Progressive within the 450VA subsidy category but regressive within the 900–6600VA subsidy range.		8% for bottom decile	12% for top decile
5.4% reduction in real income for bottom quintile	4.1% reduction in real income for top quintile	21.2% for bottom four deciles	...
...	...	...	...
...	...	...	...
...	...	...	...
...	...	...	...
...	...	...	...
...	...	...	...
...	...	...	...
Regressive		23.9% for bottom four deciles	...

(Continues on the next page.)

Distributional Incidence of Subsidies (*continued*)

Country	Energy subsector	Source	Energy expenditure as a share of household expenditure	
			Poor	Non-poor
Mexico	Electricity	Coady and others 2006	…	…
Moldova	Electricity	Tajikistan PSIA, 2006; Moldova PSIA, 2004	6.3% for bottom quintile, 4.7% for the poor	3.0% for top quintile, 3.4% for the non-poor
	Central heat	PSIA, 2006	0.1% for bottom quintile	1.8% for top quintile
	Central gas	PSIA, 2006	1.0% for bottom quintile	1.9% for top quintile
	LPG	PSIA, 2006	0.4% for bottom quintile	0.4% for top quintile
Mongolia	Heating	PSIA, 2003	10.8% for the poor on heating (18% in winter months)[b]	5.7% for the poor on heating (10.1% in winter months)[b]
Morocco	Diesel and fuel	Coady and others 2006	…	…
	Energy	Program document for DPL-II, 2007	8.7% for bottom quintile	9.3% for top quintile
Poland	Electricity	Tajikistan PSIA, 2006	5.8% for bottom quintile	2.9% for top quintile
Sri Lanka	Fuel	Coady and others 2006	…	…
Tajikistan	Electricity	PSIA, 2007	< 4% for bottom quintile	…

a. Numbers are calculated for Djibouti Ville and shown in parentheses for other towns.
b. For ger districts in Ulaanbaatar.
Note: LPG = liquefied petroleum gas.

Income impact of subsidy removal/price increase		Share of benefit from the energy subsidy	
Poor	Non-poor	Poor	Non-poor
...	...	Majority of subsidy goes to upper-middle-income households (deciles 6-8)	
1.6–5.4% increase in household expenditure for bottom quintile	1.1–3.6% increase in household expenditure for top quintile	...	...
0.0–0.1% increase in household expenditure for bottom quintile	0.7–1.8% increase in household expenditure for top quintile	...	...
0.4–1.0% increase in household expenditure for bottom quintile	0.7–1.9% increase in household expenditure for top quintile	...	...
0.2–0.4% increase in household expenditure for bottom quintile	0.1–0.4% increase in household expenditure for top quintile	...	...
...	...	15% of lifeline tariff to the poor	85% of lifeline tariff to non-poor
...	...	<10% for bottom quintile	33% for top quintile
...	...	...	...
...	...	...	...
2.9% reduction in real income for bottom quintile	2.2% reduction in real income for top quintile	25.1% for bottom four deciles	...
16% increase in spending for poor households under tariff adjustment, even with continued subsidies and lifelines.	...	...	...

Management Response

1. The Bank's energy-efficiency work in the 1990s was guided by the 1993 policy paper, *Energy Efficiency and Conservation in the Developing World: The World Bank's Role*, and by the companion "Power and Energy Efficiency—Status Report on the Bank's Policy and IFC Activities."

2. Management notes that the definitions underlying the figures it shared with IEG reflect, as well as energy efficiency captured by IEG, all World Bank lending for (i) supply-side energy-efficiency measures, including power generation plant rehabilitation, transmission and distribution loss reduction, and energy sector technical assistance with pricing covenants, and (ii) development policy lending with energy price reform.

On this basis, IEG observes that it may need to revise the language in order to describe more precisely the measures cited in the report and the differences between them. IEG acknowledges that alternative definitions of energy efficiency are possible. IEG has reported the proportion of energy efficiency projects using both stricter and broader definitions. The latter used the management-supplied information to calculate the proportion of projects incorporating plant rehabilitation and transmission and distribution measures. IEG has reported, separately, the proportion of projects involving price reform.

Chapter 1

1. A detailed exposition is beyond the scope of this report, and unnecessary given a proliferation of reviews on the subject. These include IPCC (2007b), Stern (2007), UNDP (2007), and the *Global Monitoring Report 2008* (World Bank 2008c).

Chapter 2

1. Climate Analysis Indicators Tool Version 5.0. (Washington, DC: World Resources Institute, 2008; http://cait .wri.org/). Data for 2000 includes 6 GHGs and land use change. Based on G77+ China.

2. Emissions per capita as shown in figure 2.1 can be decomposed into power-related emissions and other energy emissions. The extent of hydropower explains about half the variance in power-related emissions per capita that is not accounted for by income per capita and heating needs (Meisner 2008).

3. The negative relationship between diesel price and emissions may in part be due to subsidy-induced measurement error—but only in part. Explicit and implicit fuel subsidies do not show up as added value in GDP. In other words, the true GDP of large energy subsidizers is larger than measured. However, this does not explain the negative relationship. The largest subsidizer, Iran, would have a 20 percent higher GDP per capita if energy were priced at economic levels (ignoring general equilibrium effects). According to our regression, this means that the reference level of emissions/capita should be 20 percent higher than we have imputed. But, in fact, Iran's relative emissions are about 67 percent higher than peers at the same measured GDP. Other diesel subsidizers have lower proportions of subsidy and higher relative emissions, so this result is not being driven by measurement error.

4. The IEG review identified errors in some of these appraisals that often tend to bias results upwards, but sometimes downward. However, the figure of $1.11/kWh, from Peru, was deemed the result of best practice in analysis. Also, willingness to pay estimates do not include various ancillary benefits, including improvements in indoor air quality and facilitation of small-scale enterprise. Moreover, reported estimates for the value of off-grid electricity were much higher than for on-grid. On balance, the figures quoted here are likely to be conservative estimates of the value of electricity to the unconnected.

5. Climate Analysis Indicators Tool Version 5.0. (Washington, DC: World Resources Institute, 2008; http://cait .wri.org/).

Chapter 3

1. Indirect emissions are those "associated with off-site production of power used by the project."

2. IFC, "Lanco Amarkantak Thermal Power Plant, Environmental and Social Review Summary."

3. CO_2 Baseline Database, version 2.0 <http://www.cea.nic.in/planning/c%20and%20e/Government%20of%20India%20website.htm>

4. See <http://www.defra.gov.uk/environment/climate change/research/carboncost/index.htm>

Chapter 4

1. These figures are now out of date because of changes in world energy prices. In many cases the effective subsidy has risen.

2. Most recent data available in *World Development Indicators 2008*, except India: proportion lost in transmission or distribution or unaccounted in 2004–05, from Central Electricity Authority (Government of India, Ministry of Power, Central Electricity Authority 2006).

3. IEG is currently undertaking a comprehensive review of PSIAs.

Chapter 5

1. World Bank Progress Reports on Renewables for 2004 (retrospective to 1991), 2005, 2006, 2007 (World Bank 2005c, 2005e, 2006b; World Bank and IFC 2007). The 2004 report included project components that improved the efficiency by which energy is produced, transformed, and used; production, transportation, and distribution of steam or hot water (heat) through an interconnected network; electrical energy-efficiency improvements; and specialized entities providing energy-efficiency services. The 2005 report included end-use thermal and electric efficiency activities, power sector rehabilitation, loss reduction in transmission and distribution, and improvements in the efficiency of district heating systems. The 2006 and 2007 reports exclude transmission and distribution rehabilitation if the share of such investments cannot be clearly disaggregated from other objectives. These two reports similarly exclude Development Policy Loans unless the efficiency share can be clearly determined.

2. While this figure includes refurbishment and replacement of some district heating plants, it may exclude supply-side investments in generation. Some Bank Group–supported thermal plants might have been more efficient than plants that would otherwise have been built. It excludes lending activity related to pricing, discussed at length in the previous chapter. It is possible also that power sector unbundling or district heating privatization activities would promote supply-side efficiency not included here.

3. Efficiency Vermont 2006 Report <www.efficiencyvermont.com>

4. Based on $1.50 for a 15W CFL with a 6,000-hour lifetime; Ashok Sarkar (World Bank), presentation on "Large Scale CFL Deployment Programs," Shanghai, May 13, 2008. <http://www.energyrating.gov.au/pubs/2008-phase-out-session4-sarkar.pdf>

5. Energy-efficiency standards and labeling Information Clearinghouse. <http://www.clasponline.org/clasp.online.worldwide.php>

Chapter 6

1. The U.S. mainland was not included in the survey.

2. Flaring levels in Russia are disputed. The officially reported level for 2004 was 15 bcm. The 2007 Russian State of the Union address quoted a figure of 20 bcm. PFC Energy (2007) used a physical model of oil production, incorporating assumptions about gas-to-oil ratios, to estimate 38 bcm of flaring; they also note anecdotal accounts that some gas is vented, which has 22 times greater GHG impact than if the gas were flared.

3. Global Gas Flaring Reduction Public-Private Partnership—Expanded Update, October 2004.

4. This is consistent with CDM rules on proprietary information, but hinders external assessment of additionality determination.

5. Reported actual reductions for the first 11 months of operation were 791,325 tons CO_2e, according to monitoring report CDM0553-MR01 filed with the CDM.

6. GGFR Status Report, October 2003 <www.worldbank.org/ggfr>

7. As this volume goes to press, the GGFR reports that estimates based on remote sensing data show a 6 percent reduction in global flaring from 2006 to 2007.

Anderson, K., and W. J. McKibbin. 2000. "Reducing Coal Subsidies and Trade Barriers: Their Contribution to Greenhouse Gas Abatement." *Environment and Development Economics* 5(2000): 457–81.

APP (Asia Pacific Partnership). 2007. "Proceedings: International Workshop to Promote the Integration of Energy Efficiency in Public Procurement, July 13–14, 2007." India, New Delhi.

Azeem, Vitus Adaboo. 2005. "Ghana: Experimenting with Poverty and Social Impact Assessments." Paper commissioned by Eurodad European Network on Debt and Development, Accra, July 2005.

Bacon, R., and M. Kojima. 2006. *Coping with Higher Oil Prices.* ESMAP No. 323. Washington, DC: World Bank.

Baig, Taimur, Amine Mati, David Coady, and Joseph Ntamatungiro. 2006. *Domestic Petroleum Product Prices and Subsidies: Recent Developments and Reform Strategies.* Washington, DC: International Monetary Fund.

Baumert, K.A., T. Herzog, and J. Pershing. 2005. *Navigating the Numbers: Greenhouse Gas Data and International Climate Policy.* Washington, DC: World Resources Institute.

Blyth, W., R. Bradley, D. Bunn, C. Clarke, T. Wilson, and M. Yang. 2007. "Investment Risks under Uncertain Climate Change Policy." *Energy Policy* 35(11): 5766–73.

Bressand, F., D. Farrel, P. Haas, F. Morin, S. Nyquist, J. Remes, S. Roemer, M. Rogers, J. Rosenfeld, and J. Woetzel. 2007. *Curbing Global Energy Demand Growth: The Energy Productivity Opportunity.* San Francisco: McKinsey Global Institute.

Carey, K. 2008. "Energy Subsidies in MENA." World Bank Regional Energy Brief. http://siteresources.worldbank.org/INTMNAREGTOPENERGY/Resources/ENERGY-ENG2008AM.pdf

Castañeda, T., K. Lindert, B. de la Brière, L. Fernandez, C. Hubert, O. Larrañaga, M. Orozco, and R. Viquez. 2005. *Designing and Implementing Household Targeting Systems: Lessons from Latin America and the United States.* Washington, DC: World Bank.

Chomitz, Kenneth M., and Craig Meisner. 2008. "A Simple Benchmark for CO_2 Intensities of Economies." Background Note for *Climate Change and the World Bank Group.* IEG, Washington, DC. Photocopy.

Chomitz, K. M., P. Buys, G. de Luca, T. S. Thomas, and S. Wertz-Kanounnikoff. 2007. *At Loggerheads? Agricultural Expansion, Poverty Reduction, and Environment in the Tropical Forests.* Washington, DC: World Bank.

Clarke, L., J. Edmonds, H. Jacoby, H. Pitcher, J. Reilly, and R. Richels. 2007. *Scenarios of Greenhouse Gas Emissions and Atmospheric Concentrations.* Sub-report 2.1A of Synthesis and Assessment Product 2.1 by the U.S. Climate Change Science Program and the Subcommittee on Global Change Research. Department of Energy, Office of Biological & Environmental Research, Washington, DC.

Coady, D., and D. Newhouse. 2006. "Ghana: Evaluating the Fiscal and Social Costs of Increases in Domestic Fuel Prices." In A. Coudouel, A. S. Dani, and S. Paternostro, eds., *Poverty and Social Impact Analysis of Reforms: Lessons and Examples from Implementation.* Washington, DC: World Bank.

Coady, D., Margaret Grosh, and John Hoddinott. 2004. "Targeting Outcomes Redux." *The World Bank Research Observer* 19(1).

Coady, D., M. El-Said, R. Gillingham, K. Kpodar, P. Menas, and D. Newhouse. 2006. "The Magnitude and Distribution of Fuel Subsidies: Evidence from Bolivia, Ghana, Jordan, Mali, and Sri Lanka." IMF Working Paper WP/06/247. International Monetary Fund, Washington, DC.

Cossé, S. M. 2003. *The Energy Sector Reform and Macroeconomic Adjustment in a Transition Economy: The Case of Romania.* IMF Policy Discussion Paper No. 03/2. Washington, DC: International Monetary Fund.

Dahl, C., and C. Roman. 2004. "Energy Demand Elasticities—Fact or Fiction? A Survey Update." In *Energy, Environment and Economics in a New Era*, 24th USAEE/IAEE North American Conference, Washington, DC, July 8–10, 2004. Cleveland, OH: IAEE.

DCI (Development Consultants International Limited). 2008. "Consultancy Services for Monitoring, Verification, and Evaluation of the Compact Fluorescent Lamp (CFL) Program." Prepared for the Ministry of Mineral Development, Kampala, Uganda. Kampala.

Efficiency Vermont. 2006. "Efficiency Vermont 2006 Report." Burlington, VT: Efficiency Vermont. www .efficiencyvermont.com.

Elvidge, C. D., K. E. Baugh, D. W. Pack, C. Milesi, and E. H. Erwin. 2007. "A Twelve Year Record of National and Global Gas Flaring Volumes Estimated Using Satellite Data." Boulder, CO, NOAA National Geophysical Data Center.

ESMAP (Energy Sector Management Assistance Program). 2007. *Technical and Economic Assessment of Off-grid, Mini-grid and Grid Electrification Technologies*. Washington, DC: ESMAP and the World Bank.

———. 2004. *Nigerian LP Gas Sector Improvement Study*. Washington, DC: ESMAP and the World Bank.

———. 2003. *Turkey. Energy and Environment Review: Synthesis Report*. Washington, DC: World Bank.

———. 2002. *Bulgaria: Energy Environment Review*. Washington, DC: World Bank.

———. 2001. *Mexico: Energy Environment Review*. Washington, DC: World Bank.

———. 1999. *The Effect of a Shadow Price on Carbon Emission in the Energy Portfolio of the World Bank—A Carbon Backcasting Exercise*. ESMAP report 212/99. Washington, DC: World Bank.

ESMAP, International Energy and Development Associates, and Energy Economy and Environment Consultants. 2004a. *Environmental Issues in the Power Sector: Long-Term Impacts and Policy Options for Karnataka*. Washington, DC: Joint UNDP/World Bank Energy Sector Management Assistance Programme.

———. 2004b. *Environmental Issues in the Power Sector: Long-Term Impacts and Policy Options for Rajasthan*. Washington, DC: Joint UNDP/World Bank Energy Sector Management Assistance Programme.

GEF (Global Environment Fund). 2006a. "India Coal-Fired Generation Rehabilitation Project." World Bank, Washington, DC.

———. 2006b. *Thailand Promotion of Electrical Energy Efficiency Project*. Washington, DC: World Bank.

———. 2004. *Climate Change Program Study*. Washington, DC: GEF.

Gerner, F., B. Svensson, and S. Djumena. 2004. "A Regulatory Framework and Incentives for Gas Utilization." World Bank, Public Policy Journal Note no. 279. The World Bank Private Sector Development Vice Presidency, Washington, DC.

Gillingham, K., Richard Newell, and Karen Palmer. 2006. *Energy Efficiency Policies: A Retrospective Examination*. Washington, DC: Resources for the Future.

Government of India, Ministry of Power, Central Electricity Authority. 2006. *All India Electricity Statistics: General Review 2006*. Delhi: Central Electricity Authority.

Government of Nepal. 1997. *Nepal: Power Development Project. Sectoral Environmental Assessment*. World Development Sources, WDS 1997-2. Washington, DC: World Bank.

GTZ (German Technical Cooperation). 2007. *International Fuel Prices 2007*. 5th edition. Oschborn: GTZ http://www.gtz.de/de/dokumente/en-international-fuelprices-final2007.pdf

Hondo, H. 2005. "Life Cycle GHG Emission Analysis of Power Generation Systems: Japanese Case. *Energy* 30 (11-12): 2141–56.

Hertzmark, Donald. 2007. *Risk Assessment Methods for Power Utility Planning*. Renewable Energy Special Report 001/07, March 2007. ESMAP. Washington, DC: World Bank.

IEA (International Energy Agency). 2008a. "Energy Efficiency Policy Recommendations: In Support of the G8 Plan of Action." IEA, Paris.

———. 2008b. *Worldwide Trends in Energy Use and Efficiency: Key Insights from IEA Indicator Analysis*. Paris: OECD/IEA.

———. 2007. *World Energy Outlook 2007*. Paris: OECD/IEA.

———. 2006. *World Energy Outlook 2006*. Paris: OECD/IEA.

———. 1999. *World Energy Outlook 1999: Looking at Energy Subsidies: Getting the Prices Right*. Paris: OECD/IEA.

IEG (Independent Evaluation Group). 2008a. *Annual Review of Development Effectiveness (ARDE) 2008: Shared Global Challenges*. IEG Study Series. Washington, DC: World Bank.

———. 2008b. *Environmental Sustainability: An Evaluation of World Bank Support*. IEG Study Series. Washington, DC: World Bank.

————.2008c. *Ukraine Country Assistance Evaluation*. Washington, DC: World Bank.

————. 2008d. *The Welfare Impact of Rural Electrification: A Reassessment of the Costs and Benefits—An IEG Impact Evaluation*. IEG Study Series. Washington, DC: World Bank.

————. 2008e. *The World Bank in Georgia, 1993–2007*. Washington, DC: World Bank.

————. 2006a. *Hazards of Nature, Risks to Development: An IEG Evaluation of World Bank Assistance for Natural Disasters*. IEG Study Series. Washington, DC: World Bank.

————. 2006b. *New Renewable Energy: A Review of the World Bank's Assistance*. IEG Study Series. Washington, DC: World Bank.

————. 2005. *Romania: Country Assistance Evaluation*. Washington, DC: World Bank.

————. 2003. *Power for Development: A Review of the World Bank Group's Experience with Private Participation in the Electricity Sector*. IEG Study Series. Washington, DC: World Bank.

IFC (International Finance Corporation). 2007. "Lanco Amarkantak Thermal Power Plant, Environmental and Social Review Summary." www.ifc.org

IMF (International Monetary Fund). 2008. "Food and Fuel Prices – Recent Developments, Macroeconomic Impact, and Policy Responses." IMF, Washington, DC.

————. 2005. *Islamic Republic of Iran: 2005 Article IV Consultation—Staff Report; Staff Statement; Public Information Notice on the Executive Board Discussion; and Statement by the Executive Director for the Islamic Republic of Iran*. IMF Country Report 06/154. Washington, DC: IMF.

IPCC (Intergovernmental Panel on Climate Change). 2007a. *Climate Change 2007: Synthesis Report*. Contribution of Working Groups I, II and III to the Fourth Assessment Report of the Intergovernmental Panel on Climate Change. Geneva, Switzerland: IPCC.

————. 2007b. *Climate Change 2007: Mitigation*. Contribution of Working Group III to the Fourth Assessment Report of the Intergovernmental Panel on Climate Change, B. Metz, O.R. Davidson, P.R. Bosch, R. Dave, and L.A. Meyer, eds. Cambridge, U.K., and New York, NY: Cambridge University Press.

Ivanic, M., and W. Martin. 2008. *Implications of Reforming Energy Use Policies in the Middle East and North Africa*. Washington, DC: World Bank.

Januzzi, Gilberto De Martino 2005. "Energy Efficiency and R&D Activities in Brazil: Experiences from the Wirecharge Mechanism (1998–2004)." www.3countryee.org

Komives, K., J. Halpern, V. Foster, and Q. Wodon. 2006. "The Distributional Incidence of Residential Water and Electricity Subsidies." World Bank Policy Research Working Paper 3878, Washington, DC.

Lampietti, J. A., S. G. Banerjee, and A. Branczik. 2007. *People and Power: Electricity Sector Reforms and the Poor in Europe and Central Asia*. Washington, DC: World Bank.

Lucas, N., P. Wooders, and M. Cupit. 2003. *Egypt: Energy-Environment Review*. World Bank EEAA. Oxford: Environmental Resources Management.

McKinsey Global Institute. 2008. *The Carbon Productivity Challenge: Curbing Climate Change and Sustaining Economic Growth*. Sydney: New Media Australia, McKinsey and Company.

Meisner, Craig. 2008. "Resources, Policies, and Technology Choice in Power Generation." Background Note for *Climate Change and the World Bank Group*. IEG, Washington, DC. Photocopy.

Michaelowa, A., and P. Purohit. 2007. "Additionality Determination of Indian CDM Projects: Can Indian CDM Project Developers Outwit the CDM Executive Board?" Climate Strategies Technical Report. Discussion Paper CDM-1, Climate Strategies, London.

Morgan, T. 2007. *Energy Subsidies: Their Magnitude, How They Affect Energy Investment and Greenhouse Gas Emissions, and Prospects for Reform*. Geneva: UNFCCC Secretariat.

Nakhooda, S. 2008. *Correcting the World's Greatest Market Failure: Climate Change and the Multilateral Development Banks*. Washington, DC: World Resources Institute.

NETL (National Energy Technology Laboratory). 2007. *Cost and Performance Comparison Baseline for Fossil Energy Power Plants*. Washington, DC: NETL.

National Environmental Protection Agency of China, the State Planning Commission of China, UNDP, and the World Bank. 1994. "China: Issues and Options in Greenhouse Gas Emissions Control." World Bank Industry and Energy Division, China and Mongolia Department, East Asia and Pacific Regional Office, Washington, DC.

Norplan. 2004. *Lao PDR Hydropower Strategic Assessment*. Oslo.

PA Consulting Group. 2006. "World Bank/GGFR: Indonesia Associated Gas Survey—Screening & Economic Analysis Report (Final)." PA Consulting, Jakarta.

http://siteresources.worldbank.org/INTGGFR/Resources/indonesiaassociatedgassurvey.pdf

PFC Energy. 2007. *Using Russia's Associated Gas.* Washington, DC: World Bank.

Pillai, P. 2008. "Strengthening Policy Dialogue on Environment: Learning from Five Years of Country Environmental Analysis." World Bank Environment Department Working Paper No. 114, Washington, DC.

PNNL/ARENA-ECO. 2003. "Energy Efficiency in the Budget Sphere of Ukraine." Pacific Northwest National Laboratory and Agency for Rational Energy Use and Ecology, Washington, DC, and Kyiv, Ukraine. www.pnl.gov/aisu/pubs/14668.pdf

Saunders, M., and K. Schneider. 2000. "Removing Energy Subsidies in Developing and Transition Economies." In *Energy Markets and the New Millennium: Economic, Environment, Security of Supply,* 23rd Annual IAEE International Conference, Sydney, 7–10 June 2000. Canberra: IAEE.

Schneider, L. 2007. "Is the CDM Fulfilling its Environmental and Sustainable Development Objectives? An Evaluation of the CDM and Options for Improvements." Öko-Institut Report prepared for the World Wildlife Fund, Berlin.

Smith, T. B. 2004. "Electricity Theft: A Comparative Analysis." *Energy Policy* 32(18): 2067–76

SNC Lavalin. 2007. *Strategic Sectoral, Social and Environmental Assessment of Power Development Options in the Nile Equatorial Lakes Region.* World Bank Report No. 39199. Montreal, Quebec, and Washington, DC: SNC Lavalin and the World Bank.

South East Europe Consultants. 2005. *Development of Power Generation in South East Europe.* Washington, DC, and Belgrade: The World Bank and SEEC.

Stern, N. 2007. *The Economics of Climate Change: The Stern Review.* Cambridge, U.K.: Cambridge University Press.

Sterner, T. 2007. "Fuel Taxes: An Important Instrument for Climate Policy." *Energy Policy* 35(6): 3194–202.

UNDP (United Nations Development Program). 2007. *Human Development Report 2007/8. Fighting Climate Change: Human Solidarity in a Divided World.* New York, NY: Palgrave McMillan for the UNDP.

UNEP (United Nations Environment Program). 2003. *Energy Subsidies: Lessons Learned in Assessing Their Impact and Designing Policy Reforms.* Geneva: UNEP.

UNESCO (United Nations Educational, Scientific, and Cultural Organization). 2006. "UNESCO Workshop on Greenhouse-Gas Emissions from Freshwater Reservoirs. Statement by Workshop Participants." UNESCO, Paris, 5–6 December 2006.

UNFCCC (United Nations Framework Convention on Climate Change). 2007. "Bali Action Plan." Bonn.

van der Laar, Evert, and Harry Vreuls. 2004. *INDEEP Analysis Report 2004.* International Energy Agency. http://dsm.iea.org

Wara, Michael W. 2008. "The Performance and Potential of the Clean Development Mechanism." SSRN: http://ssrn.com/abstract=1086242

World Bank. 2008a. "Argentina—Energy Efficiency Project, Project Appraisal Document." World Bank, Washington, DC.

———. 2008b. "Development and Climate Change: A Strategic Framework for the World Bank." World Bank, Washington, DC.

———. 2008c. *Global Monitoring Report 2008—MDGs and the Environment: Agenda for Inclusive and Sustainable Development.* Washington, DC: World Bank.

———. 2008d. "Russian Federation—Kazan Municipal Development Project Performance Assessment Report." Report No. 42709. World Bank, Washington, DC.

———. 2008e. *World Development Indicators.* Washington, DC: World Bank.

———. 2007a. *Carbon Finance Unit. Annual Report 2007: Carbon Finance for Sustainable Development.* Washington, DC: World Bank.

———. 2007b. *Environmental Priorities and Poverty Reduction: A Country Environmental Analysis for Colombia.* E. Sanchez-Triana, K. Ahmed, and Y. Awe, eds. Report No. 40521. Washington, DC: World Bank.

———. 2007c. "ICR for Electrobras Energy Efficiency Project, (Loan/Credit No. 047309)." World Bank Report ICR0000184, Washington, DC.

———. 2007d. "Note on Cancelled Operations. Electrobras Energy Efficiency Project (IBRD-45140)." World Bank Report NCO0000262, Washington, DC.

———. 2007e. "Program Document for a Proposed First Infrastructure Development Policy Loan (IDPL 1) in the Amount of US$200 Million to the Republic of Indonesia." World Bank Report No. 41275-ID, Washington, DC.

———. 2006a. *Bangladesh: Country Environmental Analysis.* South Asia Environment and Social Development Unit. Report No. 36945. Washington, DC: World Bank.

———. 2006b. *Improving Lives: World Bank Group*

Progress on Renewable Energy and Energy Efficiency in Fiscal Year 2006. Washington, DC: World Bank.

———. 2006c. *India: Strengthening Institutions for Sustainable Growth. Country Environmental Analysis.* South Asia Environment and Social Development Unit. Report No. 38292. Washington, DC: World Bank.

———. 2006d. *Making the New Indonesia Work for the Poor.* Washington, DC: World Bank.

———. 2006e. *Pakistan: Strategic Country Environmental Analysis.* South Asia Environment and Social Development Unit. Report No. 36946. Washington, DC: World Bank.

———. 2005a. *Arab Republic of Egypt: Country Environmental Analysis 1992–2002.* Water, Environment, Social and Rural Development Department. Report No. 31993. Washington, DC: World Bank.

———. 2005b. *Egypt—Toward a More Effective Social Policy: Subsidies and Social Safety Net.* Social and Economic Development Group. Report No. 33550. Washington, DC: World Bank.

———. 2005c. *Progress on Renewable Energy and Energy Efficiency—Fiscal Year 2005.* The Energy and Mining Sector Board. Washington, DC: World Bank.

———. 2005d. "Program Document for a Proposed Second Development Policy Loan (DPL 2) in the Amount of US$400 Million, Annex 3: Poverty and Social Impact Analysis." World Bank Report 34439-ID, Washington, DC.

———. 2005e. *World Bank Group Progress on Renewable Energy and Energy Efficiency: 1990–2004.* Washington, DC: World Bank.

———. 2004a. *Islamic Republic of Iran: Energy-Environment Review Policy Note.* Water, Environment, Social and Rural Development Department. World Bank Report No. 29062-IR. Washington, DC: World Bank.

———. 2003. *Serbia and Montenegro: A Country Environmental Analysis.* Environmentally and Socially Sustainable Development Department. Report No. 28800. Washington, DC: World Bank.

———. 2002. *Building Blocks for a Sustainable Future: A Selected Review of Environment and Natural Resource Management in Belarus.* Environmentally and Socially Sustainable Development Department. Washington, DC: World Bank.

———. 2001a. *Making Sustainable Commitments: An Environment Strategy for the World Bank.* Washington, DC: World Bank.

———. 2001b. *The World Bank Group's Energy Program—Poverty Reduction, Sustainability and Selectivity.* Washington, DC: World Bank.

———. 2000. *Fuel for Thought: An Environmental Strategy for the Energy Sector.* Washington, DC: World Bank.

———. 1999. *Come Hell or High Water: Integrating Climate Change Vulnerability and Adaptation into Bank Work.* Environment Department Paper No. 72, Climate Change Series. Washington, DC: World Bank.

———. 1995. *Expanding the Measure of Wealth.* Washington, DC: World Bank.

———. 1993. *Energy Efficiency and Conservation in the Developing World: The World Bank's Role.* Policy Paper No.11987. Washington, DC: World Bank.

———. 1992. *World Development Report 1992: Development and the Environment.* New York: Oxford University Press for the World Bank.

World Bank HDN (Human Development Network). 2008. "Guidance for Responses from the Human Development Sectors to Rising Food and Fuel Prices." HDN, Washington, DC.

World Bank and IFC. 2007. *Catalyzing Private Investment for a Low-Carbon Economy: World Bank Group Progress on Renewable Energy and Energy Efficiency in Fiscal 2007.* World Bank: Washington, DC.

World Bank and S. E. E. C. (South East Europe Consultants Ltd.). 2005. *Development of Power Generation in South East Europe: Implications for Investments in Environmental Protection.* ESMAP. Washington, DC: World Bank.

www.ingramcontent.com/pod-product-compliance
Lightning Source LLC
Chambersburg PA
CBHW080240270326
41926CB00020B/4321